FIRE DUE TO ELECTRICITY
- PREVENTABLE

By

AJIT N. KULKARNI

First Published in 2022

Becomeshakespeare.com

One Point Six Technologies Pvt Ltd
119-123, 1st Floor, Building J2, B-Wing,
Wadala Truck Terminal, Wadala (East),
Mumbai 400022, Maharashtra, India
T: +91 8080226699

ISBN - 978-93-5458-923-2

A major fire broke out at CST railway station during the evening of June 27 2014. (Times of India photo)

To my wife Jyoti for her unceasing support...

Ajit Kulkarni,

He is a Graduate in Electrical Engineering and has a wide experience of forty years in various types of projects. He started his career as a trainee engineer in reputed organizations and thereafter ventured into business of electrical contracting. Then he shifted his attention totally to electrical consultancy, leaving contracting field. From proprietary concern to private limited company, scaling new heights, was the journey of his consultancy business. He started business as an Electrical Consultant and then formed an organization to give entire gamut of engineering consultancy.

He is the Managing Director at M/s Ajit Kulkarni Consultants Pvt. Ltd., an ISO 9001:2015 certified company, providing consultancy services to - IT offices, Data Centers, Call centers, Banks, Commercial projects, Residential complex, Industrial projects, Hospitals, Hotels etc. He has completed 1100+ projects.

The organization is offering MEPF Consultancy including Electrical, Fire, Security, HVAC, IBMS, Electrical Safety Audit, and Power Audit. They have worked for leading architects, multinational organizations, international consultants, government organizations, private companies and have many satisfied customers.

He has provided design consultancy for a wide array of clients which included multi-national companies, national companies, and even private corporate entities. His clients include brands such as Amdocs, Awaya, Axa, Atos, Bank of America, Capgemini, Capita, Citi Bank, Deloitte, Deutsche Bank, EDS, Eaton, Flextronics, Godrej, HSBC, IDFC, JM Finance, Kohler, Lks, Master card, OOMC, Principle Global, Piramal, PTC, Qlogic, Redhat, State street Bank, Standard Charted Bank, TCS, United Parcel, VM Ware, Willis, Yash, Ziemann to name a few as there are many more.

He has shown exemplary performance while doing design of fast-track IT fit out projects which are sustainable, resilient, energy conscious, and have /use latest technology.

He has received the prestigious Entrepreneur's Award from-Maharashtra Chamber of Commerce Industry & Agriculture. He has also received the Society Interiors Building Design Award and few accolades.

He has published many technical articles in journals, magazines and conducted many seminars for architects, students on technical subjects.

He has also completed many courses such as fire and safety management, EHV Substation, Solar, Data centre, Electrical Safety, façade lighting etc. And continues to keep himself updated in this field. He is member of many technical associations and institutes. A testimonial to his expertise is the fact that three patents on his name, are awaiting for approval.

He is passionate about reading technical, management, self-help books and visited worldwide to visit manufacturing facilities and for technical interactions. He regularly practices yoga and mindfulness meditation, also has affinity towards charitable institutes.

He considers work as worship and is committed to serving at the feet electricity. He assures to provide quality service, value for customer's investment, and energy efficient sustainable systems.

FOREWORD

Massive electrification in recent decades has led to an increase of the associated risks. The first thing that comes to mind is electric shock that needs off course to be considered. But we can see globally a very positive effect of standardization and education to reduce this risk. A second -and too often forgotten - risk related to electricity is the risk of fire that kills more people than electric shock and may have dramatic consequences on business continuity. When we look at electrical fire statistics, globally and in India we don't see a positive trend despite improvement of appliances, electrical equipment, and protective devices.

At IEC level electrical fire risk in electrical installation is addressed by IEC 60364-4-42 for general rules and additionally in parts 7 for special locations (see for example part 7-712 for photo-voltaic installations, part 7-705 for agricultural premises). We are now considering for future edition of part 4-42 potential improvement thanks to final circuit arc fault detector (AFDD which stands for Arc Fault Detection Device) and earth leakage detection with Residual Current Device (RCD) but also internal arc fault in electrical panel. Rising of direct current (DC) applications needs also to be considered.

Electrical device manufacturers are increasingly providing innovative solutions helping to reduce the risk of poor connection, integrating RCD and AFDD functions in protective devices, providing permanent monitoring of installation thanks to digital system.

But all these technical solutions, code & standard improvement will be nothing without the necessary awareness and education of all the stakeholders of electrical installations. That is where this amazing work done by Ajit N. Kulkarni holds a great place in this industry. Indeed, this book puts together all the practical aspects of protective measures against electric starting from basics on wiring systems and earthing arrangements, to most advanced and "active" solution like arc fault detection and that for a wide audience, including fire fighters. In addition, this book is not only providing recommendation and guidance but aiming to change the mindset of all actors to reduce the risk as far as possible. And we all know that risk reduction starts with change of behavior.

As an IEC expert and distinguished technical expert within Schneider Electric in charge of electrical fire prevention matter, I am delighted to introduce this book. I am convinced it will be useful source for all person involved in the safe and efficient implementation of low voltage installation.

Mathieu GUILLOT

Mathieu GUILLOT is a member of the IEC TC64/MT2 working group, in charge of overcurrent protection and thermal effects in the IEC 60364 series "Low voltage electrical installation" and actively prepared the future edition 4 of IEC 60364-4-42 "Protection for safety – Protection against thermal effect"

He has been working in the electrical distribution domain for Schneider Electric since 1994, where he has held various positions in R&D for HV protection, HV & LV Solution and Services and R&D for LV power circuit-breakers and enclosures. He is recognized as a distinguished technical expert in power domain by Schneider Electric internal recognition program "Edison". He is the author of numerous technical guides for Schneider Electric. In 2019 he was appointed electrical fire prevention "Subject Matter Expert" and published two papers/articles: *"Electrical fire prevention - Discover how to mitigate risk of fire for new and existing commercial buildings"* – 2019, *"New gas and particle sensing technology detects cables overheating in LV equipment, 2021"*

PREFACE

'If we could change ourselves, the tendencies in the world would also change. As a man changes his own nature, so does the attitude of the world change towards him. We need not wait to see what others do.' Mahatma Gandhi

Every now and then we hear news of fires and in most of the cases electrical short circuits is the reason behind it. In fire not only life and property are lost but along with that there is loss of intangible things, like dreams, passion, association, labor, planning, business, assets which turns in to mere ashes.

There could be dream of purchasing a house or shelter for which all life earnings or part is put in to it. There could be passion like designing of building or office premises or commercial complex or hotels or any other thing. There could be detailed planning by architects, planners or engineers. There could be contribution for building the project by tradesmen or labor. There could be construction which would strengthen nation building. There could be ray of hope like conquering pandemic. All this tangible or nonphysical things turn in to vestiges.

As per record of US in 2015 there was loss of 14.3 billion dollars. In India, every year nearly Rs. 1000 Crores lost and 10000 lives are lost. Nearly 75% of the fires are due to electricity and we have to change this situation.

Loss due to Fire is quite disheartening and that too fire due to electricity is very painful, since it is preventable. When I read "fire was due to electricity or short circuit was the cause of fire" it hurts me more since I am an electrical design engineer and a professional consultant. I understand that cause of fire could be electricity but not because electricity is hazardous but because it was mishandled. Electricity is not bad but handled badly.

The actual nuances of reasons behind electrical fire are hardly analysed, all evidence lost in fire may be one of the genuine reasons for this but also, I feel mainly it may be the ignorance and unawareness about proper workmanship and education about the same. At this point, I feel now is the time for me to share my knowledge and experience in this field with a goal towards zero fires due to electricity. With a desire to contribute for betterment of society by educating people about technical lapses, dos and don'ts. I have made this humble effort of putting all the information together/ of writing a book

Attempt is also made here to form a knowledge bridge between fire and electricity which are presently isolated apart as if right hand does not know what left hand is doing.

If we see strength and weakness during execution of project, our strengths are Indian standards of products, methods of installations, national building codes, Indian electricity act and CEAR rule, local norms and our design knowledge etc. Knowledge being the continuous process of improvement, it cannot stop at any level. As such theory part is strong. But when it comes to practical implementation, although we have successfully completed many projects, at grass root level during installations or in testing some mistakes do happen, mostly unintentionally. Although standards are there, inadvertently compromises are made and their lies the issue.

From this point of view all efforts are made here, to highlight fundamentals of fire and electricity and more about "Electrical Fire". The topics covered are explained mostly by giving practical examples.

If they are thoroughly understood and implemented wholeheartedly, then electrical fire incidences can be averted in daily life or in professional life. This achievement will automatically lead us towards our target, nil fires due to electricity.

ACKNOWLEDGEMENT

My projects- While interacting with clients, project managers, architects, suppliers, fellow consultants, contractors, supervisors, workmen in each project there is always some new learning. Projects have always triggered the sharing of knowledge and learning. Suppliers have given technical data whenever required and arranged technical trainings on products, standards, new developments.

Mathieu Guillot – During his busy schedule he carved out time to go through the book and gave valuable suggestions to incorporate additions and changes. He has given forward for the book. As such sincere thanks for the same.

My Office- During my absense in daily work, all my engineers, draftsmen, admin staff have worked very diligently and kept business running.

My family- Jyoti, Amogh, Mandeep, Omkar, Janhavi, Tanisha, Ishita tolerated me while working on this writing project. Amogh has helped in making drawings, giving references required along with doing routine office work.

Publishers- BecomeShakespeare have helped in getting this content published.

Ajit N. Kulkarni

PROLOGUE

Electrical fire is mainly related to two streams; one 'Electrical' and the other 'Fire'. Fire and Electricity go hand in hand in any installation but their remains a disconnect in two subjects. People working in respective areas may be proficient in their own field but may have little knowledge about the other. For e.g. firemen may not be aware why, how and where electrical fires may erupt? Whereas electricians may not know about different classes of fire and how to douse them at the preliminary level. Idea of this writing is to remove this disconnect and to form a bridge guided by knowledge between the two.

Another purpose is to polish and fine tune the practical working knowledge of people in electrical field. So that they will realize the importance of implementation of good working practices and consequences of negligence at work and the warning signals for it.

Taking into consideration benchmark of good design, planning, applying codes and standards, I contemplated and visualized about the mistakes that frequently occur or could occur mainly at execution level and this is the focal point of explanation in this book.

The purpose of writing this book is also to revisit the <u>basics of fire, till safety part associated</u> with it. So, concept of 'EFACTS' is designed and taught by me. They are 'Electrical Fire', Awareness, Causes, Testing and Safety. All these are discussed and elaborated in further chapters.

There are people with different backgrounds in the field of electrical & fire. They implement their experience and knowledge in their routine work in both the fields. They are-

Category 1 – People who have proficiency in fire but are not aware of electrical intricacies.

Category 2 - They are absolutely ground level people who work with their hands like lone electricians, electricians attached to shops, small contractors who are far away from standards, rules regulations etc. but have knowledge of trade because of basic level of education or hands on experience.

Category 3 – Knowledgeable and experienced contractors, engineers who have been working for years. They know standards, codes, rules etc. but are slightly moved away from theory, practical mistakes leading to improper installations, detailed testing, site issues etc.

Category 4 – Intellectuals having partial exposure to both the topics and also have a say and last decision in projects such as architects, planners, project management consultants, clients, government officials, fire brigade personnel etc.

All such people are target audience/readers and beneficiaries who will benefit from this knowledge and experience-based information.

For the aim 'not to have fire due to electricity', good exposure to electricity and electrical field is required. And hence essentials of this stream are explained at length. At the same time in case of fire, issues like fire theory, information of actual incidences of bursting of fire and how to extinguish at preliminary level is also important. So accordingly, fire topics are introduced. In both the topics concise but crucial approach is kept in mind.

This attempt is information sharing which can be used by all the above people and if they implement it in practice all of us can overcome this electrical fire issue.

Contents

THOUGHT PROCESS

Mindset

"Whatever the mind can conceive and believe, it can achieve"

Napoleon Hill

Worldwide all are concerned with electrical fire and prevention of fire is largely discussed in many forums, conclaves, seminars, standards, books etc. and by all government organizations, non-government organizations, associations etc. Every individual who is associated with such activities are making all out efforts in different ways to educate the users, electricians, engineers, organizations. When we are talking about fire due to electricity then all electricity users are covered under this umbrella of protection. Despite all out efforts there are failures either in electrical installations, maintenance or design or in material and fire finds the way out and endangering life and material. What could be reason? Let us investigate different way.

'Change the mindset of individual'.

Who is individual? One who is User or one who is Provider of the electricity.

'User' of electricity means any one out of large group of people/agencies such as who are using electricity as 'end user'. Those are-

- Residence
- Industries
- Commercial complex
- Office environment
- Hotels
- Hospitals
- Educational institutes
- Agricultural
- Infrastructure
- Railways
- Marine etc.

'Electricity provider' means any one out of large group of people / agencies who are directly or indirectly involved in providing electricity such as

- Electrical supply companies
- Distribution networking organizations
- Planning departments

- Contractors

- Consultants

- Project Managers

- Suppliers

- Engineers

- Electricians

- Workmen

- Maintenance staff

- Facility management staff etc.

Everyone who is linked in chain of user as well as of provider must change the mindset. One must think in their mind while handling electricity that 'Whatever I am doing, is it safe or is it dangerous and can aggravate chances of fire'? Mindset should not be like : Because someone is telling so I am doing it or What's the harm when I am doing it for years ? or I have to fit in budget or I have to complete the work in certain time or could be any other reason. Primarily thought should be; are there chances of electrical fire with whatever I am doing?

Every individual must think that it is 'my' responsibility to handle electricity safely so that danger can be averted. One should not stop at 'individual' level but must go beyond 'self' and guide others and 'light the mind' if someone else is making mistake.

In between 'users' and 'providers' there lies 'work' which is sandwiched between two-

- Conceptualization
- Planning
- Concept notes
- Design Basis Reports
- Design
- Design Drawings
- Estimation
- Tendering
- Negotiations
- Ordering
- Site Movement
- Construction drawing
- Installation
- Changes
- Site testing
- Factory testing
- Quality checks
- Pre-commissioning checks
- Testing
- Certification and approval

- Commissioning
- Handing over
- Maintenance
- Modifications
- Rechecking
- Recertifying

One may realize that quality of sandwich will vary when quality of each factor will vary. So, there needs to have harmony among the three factors- users, providers and works. If all three, functions as per prescribed rules and implement good working practices, then it will be healthy installation and there will not be any untoward incidence of electrical fire, which is the aim.

EFACTS

In electrical fire, ignition happens due to many reasons and if it comes in contact with combustible materials then it turns in to Fire.

Let us understand the concept of EFACTS. It is understanding of –

- Electrical fire
- Awareness
- Causes
- Tests
- Safety

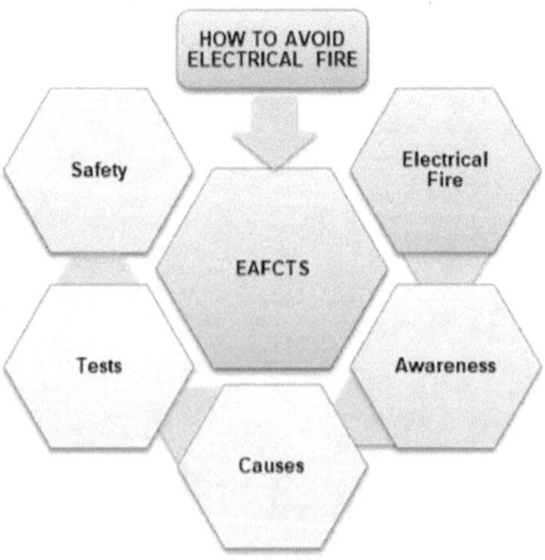

Electrical Fires – Primarily one must be able to distinguish between (i) fire occurring due to electricity and (ii) fire due to other causes except electricity. Afterwards it is easy to understand fire due to electrical installation.

Awareness- Many issues related to fire will surface and raise the alarm if there is awareness about Electrical installations. Simple mindfulness awareness can avoid electrical problems and will prevent fires. Wrong thing which can be seen, smell, hear and even feel can be dealt with Personal intervention. This initiative will avoid many fires.

Causes- Improper installations can be understood by technical / knowledgeable person. They can then visualize wrong installation and raise the alarm. Action can be then initiated to avoid the fire incidents. Examples of inappropriate installations and likely causes are given in following chapters. Understanding them will help to check the reasons and avert fire.

Tests- Thorough testing for any installation, big or small is necessary. When pre-commissioning checks will be done strictly, all required testing done properly, then only installation can be further taken up for commissioning. These tests are to be repeated on regular basis yearly till installation is present. This will ensure more robust and healthy fireproof installation.

Safety- At the time of installations and thereafter, strict adherence to safety precautions is necessary. Thus, fire incidences due to unsafe methods of erection and working can be avoided.

ELECTRICAL FIRES
Ignition Sources Theory

As all of us know that due to electricity ignition happens which further escalates to Fire. It will be worthwhile to understand theoretical explanation of ignition due to electricity first and then to move to practical scenario. In case electrical equipment such as cables, wires, switchgears, conductors etc. which develop energy, comes in contact with combustible materials then combustible materials ignite and fire occurs or starts.

Electrical ignition sources which can start ignitions can be subdivided in to –

1. Voltaic Arc Phenomena

2. Resistance heating and

3. External heat source

There could be either of reasons or could be multiple reasons which are sufficient to ignite.

1. **Arc**

    Arc is a luminous high-temperature electric discharge across a gap or a joint. This is also said to be voltaic arc.

The temperature in an arc can be several thousand degrees, depending on current, voltage and the type of metal. In given circuit if insulation fails then arcing can happen. It may be in series, parallel or with ground depending upon how arcing happens.

If there is breakage in conductor like splicing or manufacturing defect, and arcing happens then, it is said as Series arcing since it is in conductor path.

If arcing happens between two conductors or multiple conductors then it can be termed as parallel arcing. Sub part to above can be parallel arcing between live conductor and ground. This can be shown as given bellow.

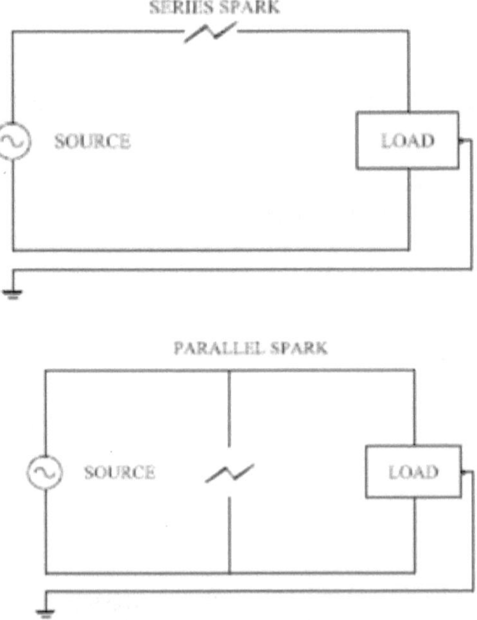

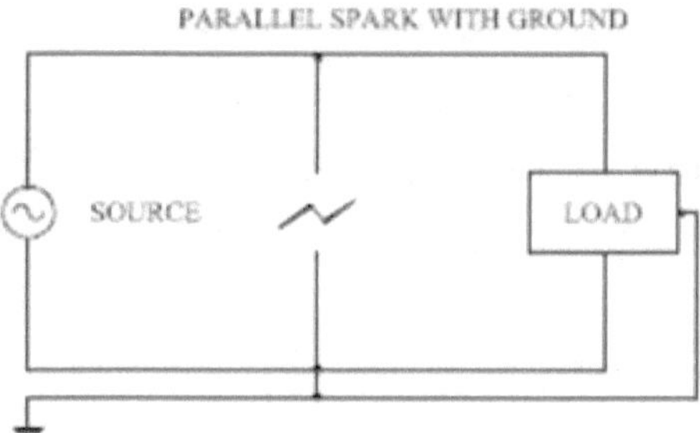

Due to arcing following indirect effects also take place -

1. Carbonization of insulation

2. Externally induced ionization of air (created by flames or an earlier arc)

3. Short circuits

CARBONIZATION OF INSULATION

Initially when arcing in circuit materializes, temperature of insulation which is a PVC material, increases. When PVC is exposed to temperatures of 200 – 300°C, it chars and the char is a semiconductor. Because of semiconductor properties, it leads to leakage currents. Once leakage currents form, then further arcing starts. Even when arcing is small and PVC is getting heated to low temperatures, and at that time if circuit voltage gets applied, then it can cause ignition of insulation.

IONIZATION OF AIR

The dielectric strength of air is high but breakdown can occur at much lower values if the air space is ionized by flames and or pre-existing arcs. Ionized gases will be ejected if fault occurs. These gases can travel to a certain distance, and if they encounter another circuit, they can readily cause a breakdown and new arcing at the second location will start. So also, Arc can initiate a fire or devastating explosion if a combustible gas/air mixture is located at the position of the spark.

SHORT CIRCUITS

In any circuit, if suddenly low resistance develops and due to available fault level, high current is produced, then that high current takes the path of low resistance. Sudden reduction in resistance can happen in healthy circuit when current carrying conductor touches ground or between each other. This can happen due to fault in circuit. It is then called as short circuit. Short circuit happens in two ways-

1. Typically, when insulation of conductor fails and touches metal casing which is grounded, then short circuit current gets established between live conductor and ground. This short circuit is high when all three phases and neutral is getting grounded. Switchgear associated to it is supposed to operate and cut off the supply. If not, then arc is getting established and short circuit chain starts.

2. Formation of arc when two or three live conductors come in contact of each other.

In all above cases fire starts due to –

- Arc getting in touch with combustible material

- Radiated heat generated due to arc, gets in contact with combustible materials

- Arc causing metal turning hot and therefore getting in contact with combustible materials

In an electrical a short circuit two main types of short circuits are present a dead short circuit and a limited short circuit.

- **Dead Short circuit**

  A dead short occurs when a live wire comes in contact with each other or with ground wire in AC circuit or the positive and negative wires are connected to each other in a DC circuit. Then the circuit is subsequently energized. In properly protected circuits this will cause the fuse to blow or protective switchgear to trip and the circuit to de-energize. This type of situation does not create sufficient heat to ignite combustibles. However, it is possible that the circuit is not adequately and properly protected. If this occurs, the current can continue to pass through the wires causing them to significantly overheat. This type of situation can ignite surrounding combustibles causing a fire.

- **Limited Short Circuit**

  The other type of short circuit is a limited short circuit. In this case, wires come in contact such that the volume of material through which the current flows is smaller than the protective switchgear. This will create a spark or flash and result in melting of the metal of the wiring. Characteristic is beading of the metal wire which is normally observed. This situation can also cause ignition of combustibles, provided the mass of the combustibles, contacting the heat source is small enough so that the heat source can cause it to reach ignition temperatures and initiate a self-sustaining exothermic oxidation reaction or fire. It is difficult to ignite concentrated, solid combustibles such as wood, plastic and even paper with this type of heat source. However, cotton products, sawdust, wood chips and combustible gases can be ignited.

2. **Resistance heating**

The causes of excessive resistance heating can be subdivided into:

1) Over load

2) Excessive thermal insulation

3) Leakage current and earth fault

4) Overvoltage

5) Poor connections

- **Overload-**

Cables/wires/Busbars/Switchgear are manufactured as per standards and have specified current rating. Accordingly, for any given circuit one must select and size the equipment such as cables, wires etc. properly. If more load is connected in circuit, then naturally more current in cables wires, equipment etc. will start flowing than specified value. In any circuit, appropriate protective switchgear is supposed to be provided so that as soon as excess current starts flowing, switchgear is supposed trip the circuit and will save the wires, cables, equipment and therefore the installation. But in absence / improper protective switchgear if circuit starts drawing more current then it will cause cables, wires, equipment getting heated and hence melting of insulation. This may be further affecting in phase-to-phase fault or phase to ground fault. Afterwards there will be arc development. If the said arc gets in touch with combustible material or if heat is produced due to arcing and gets in touch with combustible material then there will be fire.

This can happen in case of direct over loading or indirect overloading. Direct overloading means in a given circuit load value is increased instead of predetermined value. Indirect over loading means, addition of increased load on small power strip or in big way like in panels.

- **Too much insulation-**

  Laboratory tests have shown that in case of cables, if more number of coils are placed one above another or if too much insulation around cable is placed as cover then cables are generating heat and because of that combustible material may easily get ignited.

- **Leakage current and earth fault-**

  In residential, commercial and industrial or in any type of installation when electricity is used as media to operate the equipment then there will be wires, cables, terminals, coils etc which will be insulated. If this insulation breaks or fails, then naturally current will start flowing towards least resistance path. Least resistance path will be next available conductor which can be casing, cover, envelope etc. If the magnitude of this current is less, then it will be leakage current. But if it is more then it will be earth fault. As a standard practice, as per rule, all electrically operated equipment has to be earthed. It means outer casing must be earthed. So, if any leakage current forms then operator or the person who is handling the equipment will not get electric shock and will remain protected. So also, switchgear selected such as earth leakage circuit breaker or earth fault relay as the case may be, can sense the leakage current or earth fault current, and will disconnect the power supply thus ensuring no future damage. However, if all these protections are not in place then leakage current or earth fault will not get earthed and will pass through least resistance path causing spark or local heating of metal or further damaging insulation. This in turn will generate heat or indirect excess heating.

Fires can also be caused when water comes in contact with electricity through what is called leakage current. Exposed wiring, which exists primarily at connectors and switches, can come in contact with water. Since water conducts electricity, a current will flow through the water between contacts or from the live to ground or phase. Over time, the water will accumulate salts which increase its ability to conduct a current. This current can eventually develop to a point where it generates a significant quantity of heat which begins pyrolysis and carbonize the combustibles in the area. This results in a situation where a carbon bridge is formed, creating a continuous arc or significant generation of heat. Ignition of surrounding combustibles can lead to a fire. Electrical boxes which are damp or wet, fire is generated as mentioned above.

- **Over voltage-**

It is well known fact that during lightning, thunder cloud discharges, very high current in large magnitude is developed. It has to be diverted to earth by lightning conductor. Due to lightning and high current strikes, voltage rises sharply in ground near vicinity. Lightning will strike directly on the buildings which have / do not have lightning protection installation. Same can happen on high-voltage overhead line or low-voltage overhead line. Lightning strike leads to formation of coupling of voltage surges, causing induction on conductor loops which are located in the vicinity of a conductor in which lightning current is flowing. It can also have indirect impact because of rise in voltage gradient when lightning strikes and passes in ground or comes in direct contact to the materials which is getting grounded. All these incidences raise the voltage level in protective conductor or in equipment or in distribution conductor. If the protective conductor is not of adequate size or if installation is nonstandard, then it causes heating in conductor. So also, if other conductive material comes in touch with this, then flash over can happen. Apart from protective conductor if the electrical distribution is not protected by voltage surge suppressor then high voltage effect is going to happen in distribution circuit. Due to high voltage, there is component burning in electronic circuitry or heating up of conductor / insulation if not protected well.

Apart from lightning, if high voltage develops due to neutral isolation or due to accidental touching of high voltage conductor to low voltage conductor, then voltage in low voltage conductor is going to rise. This will cause burning of components like TV, coils, chokes etc. and more and more heat will be generated.

- **Poor Connection/Contact failure-**

Contact failure is by and large referred over as (1) Connection of wires/ cables to connector. (2) Interconnection of live conductors like bus bars or conductors. Connection points in electrical installations are the weak spots in the system unless connections are properly done. Fire due to errors can break out due to connection failure in such junction point. Weak spots are where cables/wires connected to terminal blocks, switchgear, socket outlets, switches. Once there is a loose contact, high resistance gets developed and over the time it will result in local heating. This developed heat will get transferred to wire/cable connected to it, raising temperature of wire/cable. It may generate spark also.

Most fire incidences show that primary mechanisms for causing electrical fires are poor connections and arc tracking. If a connection is not mechanically tight and of low resistance, it can start to undergo a progressive failure. The process often has the property of instability that means fluctuations and has positive-feedback loop. High resistance creates localized heating, leading to increased oxidation and creep. Then connection becomes less tight, and further heating occurs, until high temperatures are attained. At a certain stage, a poor connection can become a glowing connection which shows very high temperatures. At that point, nearby combustibles may be subject to ignition.

Other reasons to support Ignition can be further detailed in following points-

Improper Materials-

Termination comprises of combination of materials, skilled labor and proper tools. Materials required are lugs, inhibiting compound, gland, flat washers, spring washers and insulation tapes. Other accessories which are required depending upon type of termination/joint are stress relief cone, ferrules, hardeners, protective jacket etc. In short depending upon type of termination or joint, different prescribed materials are required. In case, if short cuts taken or non-standard materials are used then joint is bound to fail immediately or if not immediately then subsequently afterwards as time proceeds.

Skilled technician –

Skilled, experience and well-trained technicians are essential requirements. Inadvertent mistakes can cause loose connections.

Tools Tackles –

Proper tools tackles like cutter, pliers, spanners, torque wrench, splicers etc. are essential requirements. Short cut taken can cause loose connection.

Environmental Protection –

Once termination is done, the usage will decide type of protection level. Ingress of water, dirt, dust or corrosive gases may cause oxidation and further will loosen connection. This will lead to damage to termination and connections will be loosen further. This will cause local heating which can promote oxidation and sagging of metal causing contact glow.

Inadequate Tightening – Adequate pressure on joint is essential. It should not be more and it should not be less. Same is essential for correct contact. If work is done in hurry then every likely contact will remain loose.

3. **External Heat source**

Ignition materializes indirectly and electrical equipment act as heat source externally when they are hot or exploded. This happens if following equipment emits the heat and combustible materials come in contact with this heat source, then combustible materials catch the fire and fire termed as electrical fire.

Such equipment are - Room heaters, chokes, warm wires / cables, Panels, Distribution boards, feeder pillars which are hot from outside CT, PT, Transformer etc.

Space heaters are major cause of electrical fires because these types of heaters are portable type and many a times people put it near combustible surfaces such as curtains, beds, clothing, couches, rugs, coil etc. Space heaters are especially dangerous in this regard because the coils become so hot, they will almost instantaneously ignite any nearby flammable surface. If you want to use space heaters, use the radiator type that diffuses heat over the surface of the appliances. These are less likely to ignite flammable items but should still be kept away from flammable materials. Hot iron on clothes can also endanger the situation in this type.

Above all discussions and theory throws the light on electrical ignition source theory. To start fire starting point is first ignition and if the same is due to electricity then it is being termed as electrical fire. But such incidences can be averted by ensuring to take care of above facts. This can also be done at by taking due care of installation while design, installation and maintenance.

Up till now lot of theory is being discussed above and now it is turn to see practicality of the same.

AWARENESS
Prevention By Remaining Aware

Best creation on this earth is human being which has many organs and brain to analyze the action which are to be carried out. Many a times we as human being do not act on the incidences what we see. We just consider as usual routine thing, ignore and fails due to do simple inaction and then suffer. In the context of Fire sometimes same happen.

When we are being told, educated, read 'n' number of times that extension board is not to be used unless one is very sure that equipment connected to it, will not overload the wiring. Still, when we see this situation, we 'just' ignore. This may cause overheating and resulting in fire. Similarly, there should not be loose connection in socket or wires are not to be put in socket without plug top. Although it has been told many a times, same mistakes are repeated and ultimately, we land in fire incidences.

So, point is 'the Sense Organs or may be said as '*Dnyanendriya*' which are gifted to human being are to be used effectively by human being so the situation which may arise at nascent stage of fire can be avoided.

One can use 'Eyes to See', 'Nose to Smell', 'Hands to Sense' and 'Ear to Listen'. And very easily one may notice or smell or sense or hear danger. Further action or alertness might be useful to control the situation and can avert the fire. In fact, there should be reflective action of raising alarm immediately after seeing, smelling, sensing or hearing the danger. Now it's time to take action to prevent the fire. How it can be done?

See- Following electrical fire prone situations are to be sensed by 'vision'. E.g. spark, loose connection, substandard / faulty installation etc. If these situations are recognized as 'dangerous' then half the battle is won. Take proper action if handy or report to concerned authorities for taking action and result will be fire avert. So just tune in your sense of seeing.

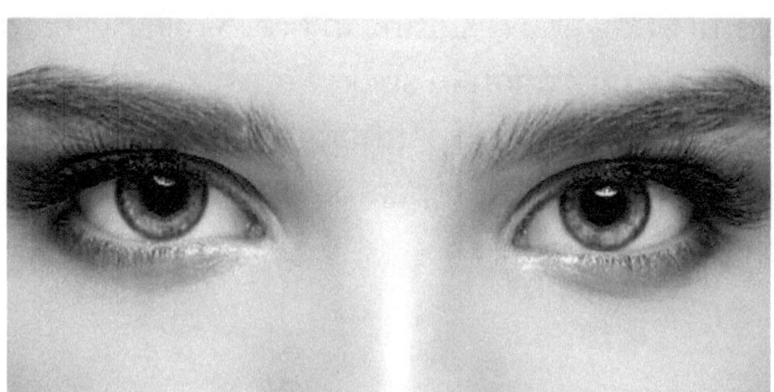

Smell – When smoke starts coming out of flammable materials or burning starts, strong odor can be sensed. If it is wire or cable then predominantly different burning smell is recognized. Immediately then raise the alarm or put off the switch to stop electrical power and you can control further burning and fire. Example is mainly in electrical wiring or in electrical panel boards, distribution boards, light fittings, switch boards etc. where burning smell starts coming well in advance. So just tune in your sense of smelling.

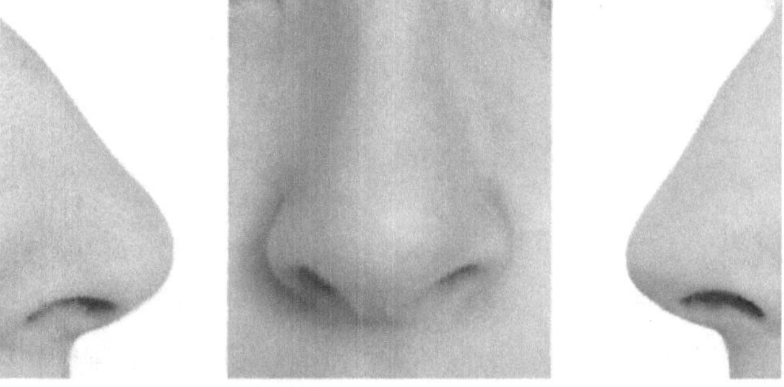

Sense – Hotness or coldness of equipment can be sensed by touch. This has to be done by knowledgeable person only and do not touch live or dead part purposefully. But accidently if a person gets touched and comes to know that equipment, wire, distribution boards are warm or hot, then they should realize that there is danger. Immediately call a concerned person or simply switch off the supply. This is usual practice in industry and technicians can cultivate habit to sense warmth, which possibly may avoid fire. But ensure that touching is limited to insulation only. And not to touch live part or conductor else will have fatal electrical shock. Please do not put life in danger. Checking of cable or wire temperature can be done by person who has license to do and not by any other person.

There are infra-red thermometers to check the temperature. One can use the same once in a week in bigger industry to see if all is proper. Infra-red thermal camera can be deployed once in six months in industry to ensure no overheating. But surely touch can give feel of basic warming action in equipment which further may lead to fire. So just tune in your sense of touching.

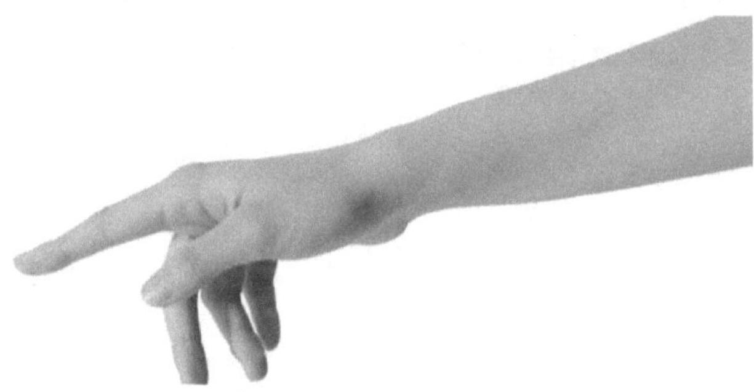

Hear- In electrical equipment there is sound of crackling which generally, happens when insulation starts breaking. Expert and experienced people can easily hear such sound but inexperienced person can at least raise alarm when such sound is being heard. Insulation breaking in any equipment or failure of rotating part in machine will subsequently lead to dangerous situation. So just tune in your sense of hearing.

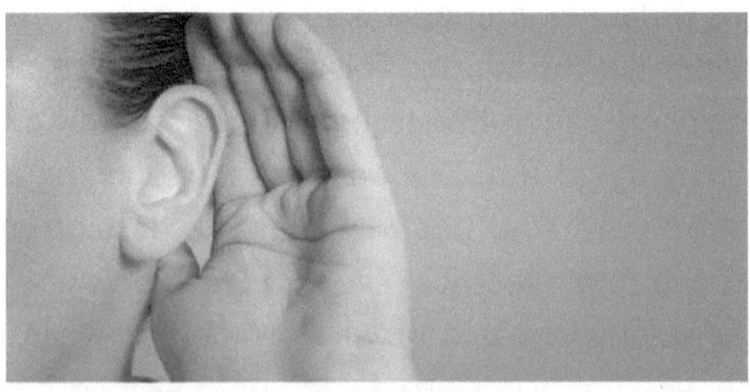

Crux of the point is that if you come across such situations accidently also or in normal life then you must use effectively your God given gift, your sense organs. Around us there are many incidences which remain unnoticed or action in such cases are not taken. Simple alertness can bring the fire situation under control. It does not require much technical knowledge but what is needed is commonsense and action.

If a person comes across a vulnerable situation leading to fire or his senses indicate burning smell or heat or see a spark or unhealthy installation then that person can raise the alarm, wherever present at residence, office, hotels, hospitals or any public places. He / She can contact concern person or authorities or owner. One must take onus on self for action, train their organs and remove fear from mind for the consequences.

Given below are few examples which will help train his/her senses to recognize the danger. Trained and alert minds will avoid inaction in such situation and avert mishaps.

1		Flexible cables are used above and below carpet without mechanical protection. Switch socket box used above carpet. One spark through loose connection and flammable carpet in marriage hall will be on fire.
2		Loose connections, improper installations. One spark and will catch fire and to add fuel vehicles with flammable fuel is available nearby.
3		Above ceiling PVC conduits used instead of MS conduits which is not as per Standard.
4		Cable shafts not closed by fire sealant so fire may pass from one floor to another.

5	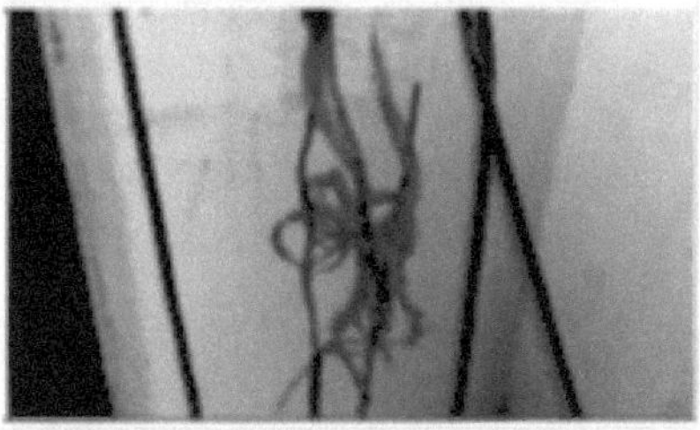	Loose wiring can call for spark
6		Construction site mess- in spite of good bus duct arrangement, loose cabling making installation hazardous.
7		Brick placed on live panel making panel at risk.
8		Live wire touching to live dry type transformer which is risky.

9	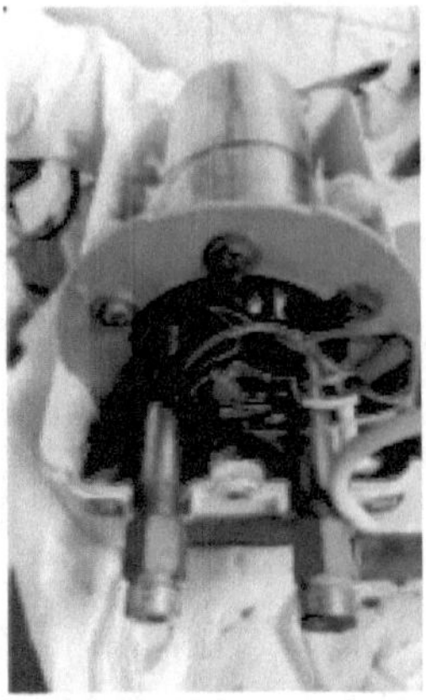	Improper wiring in geyser which will get heat up due to hot water and further insulation damage and risk.
10		Live panel and nearby flammable paint increasing fire load, cleanliness in electrical room is required.
11	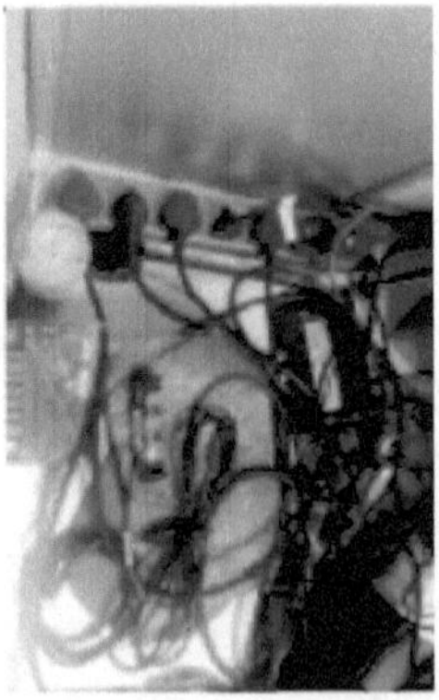	Classic example of wiring mess, proper dressing of wires is a must.

12		Hanging wires without support calling for risk of loose connection.
13	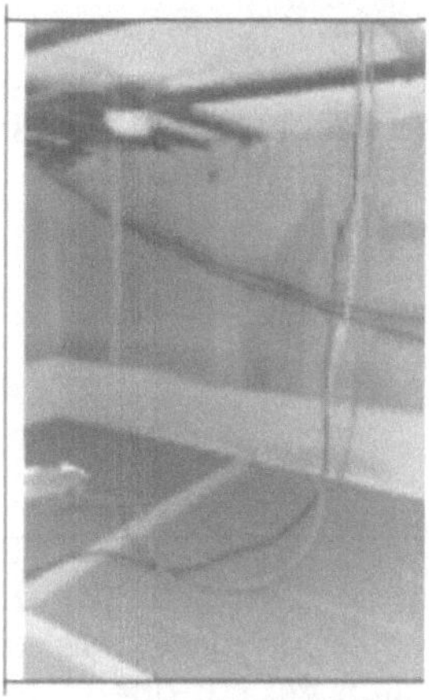	Improper installation above false ceiling, loose wire without conduit for light fitting.
14	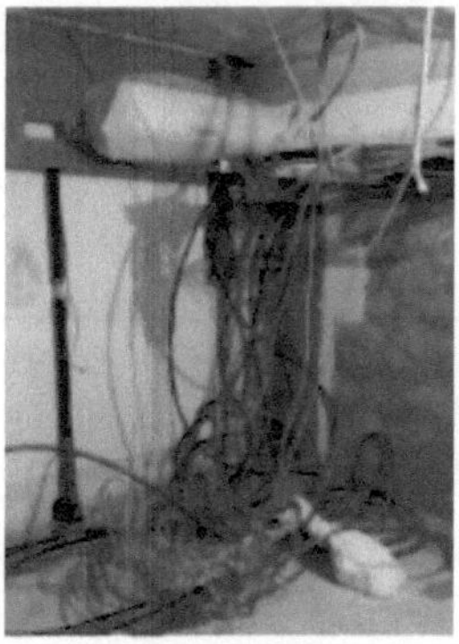	Construction site wiring mess will call for nothing but fire.

15	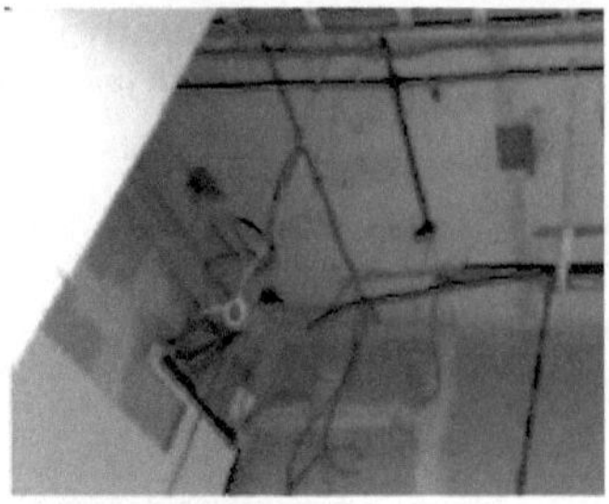	Improper wiring above false ceiling without conduit and support.
16		Charged panel but cable chamber door is open and also cable coming out of cable chamber.
17		Insulation of flexible cables might get cut due to sharp edge. So proper bushing required.
18		Indoor substation used as store, not right practice.

19		Battery rack without cover from mechanical damage.
20		AC unit above dry type transformer is improper.
21		No clearances available which is risky.
22	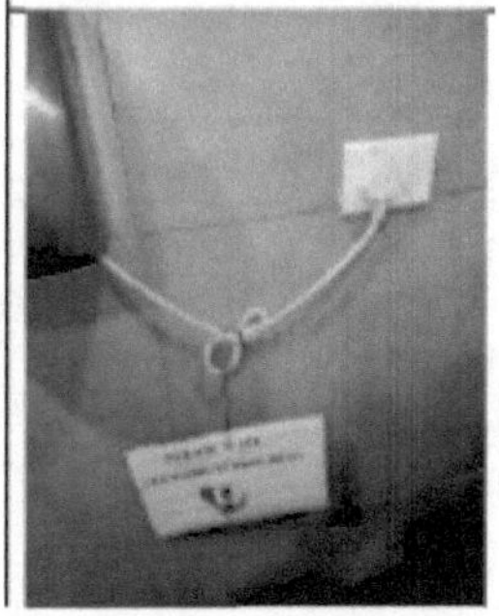	Mockery of electrical installation and that too metallic hook on wire means chances of cutting insulation.

23	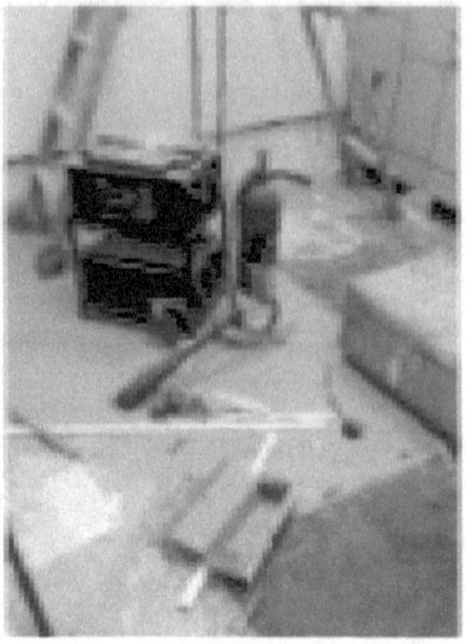	Fire hazard due to open batteries, installation accessories and extinguisher next to battery so in case of fire it cannot be operated.
24		No proper place for distribution board, which is under staircase along with utensils making fire hazard.
25		Improper method of welding with all flammable material around which is very dangerous.

26	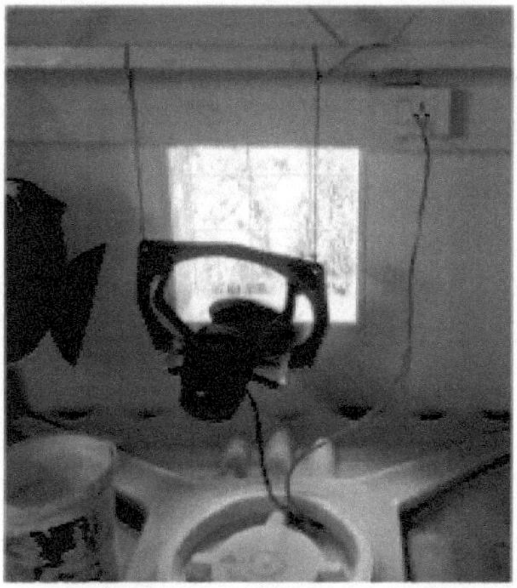	Without plug top wiring with improper installation of exhaust fan, flammable paint nearby dangerous work.
27		Without plug top wiring, prone for spark.
28		Exhibition scene with loose without support wiring and fan is making vibration into wiring making chances of loose connection and sparking.

29	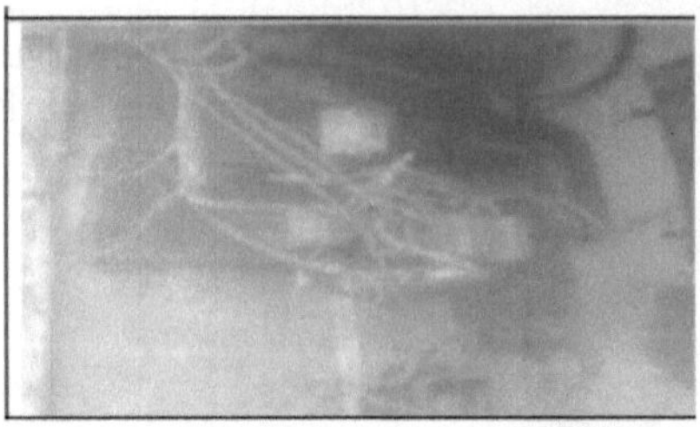	Mess is meter box which will call for nothing but electrical fire.
30		Metering kiosk bad dangerous installation.
31	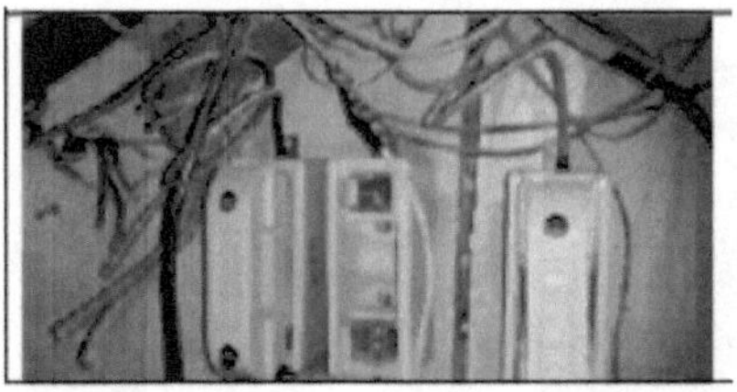	Bypass fuse and improper wiring will cause for danger.
32		Such bad installation never to do.

33		Loose connection made local burning and insulation damage and start of fire.
34		Glanding is must and cabling to be installed in straight manner not in circle making installation loose connection and hazard.
35		No lugs are used and in one terminal multiple wires installed making installation hazardous and already burning process started.
36		Never to bring water in contact with electricity else electrocution.

| 37 | 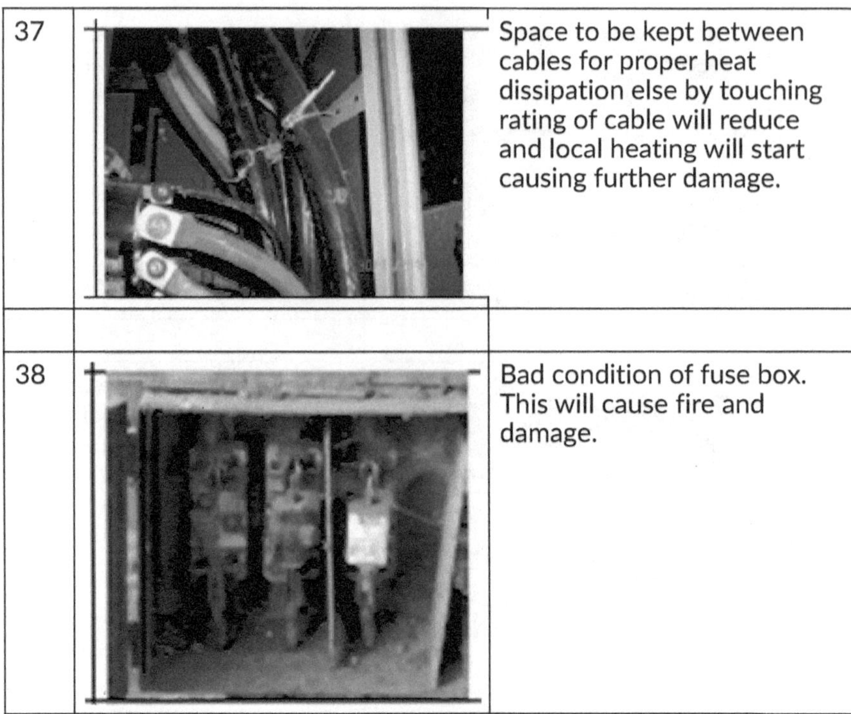 | Space to be kept between cables for proper heat dissipation else by touching rating of cable will reduce and local heating will start causing further damage. |
| 38 | | Bad condition of fuse box. This will cause fire and damage. |

Images by Author

People should take a note of such faulty situations and take corrective action immediately. If the same is ignored then there might be direct fire or installation will get deteriorated slowly and after some time situation will turn into fire.

It is recommended to have periodic visit by maintenance staff to electrical room and electrical installation to 'See, Smell, Hear and Sense' to note the warning signs of fire. The frequency can be decided daily, weekly, monthly or yearly depending upon type of installation. Same is being elaborated in 'frequency of testing'.

CAUSES
Examples of Electrical Ignition

So far, we have learnt the theory about electrical fire. We have also seen reasons and practical examples of faulty installations and recognized that alertness and proper action is required to avert the fire.

Now we shall see the important factors and likely causes of electrical fire and ways of prevention.

When fire takes place, all evidences are burnt and real reason remain unseen and unknown except for short circuit. But information is always collected for likely causes especially in small fires which were controlled and disasters were avoided.

Taking a review of all past experiences, root causes are highlighted here. The measures to avoid such situation, the reasoning behind them is also explained. This will help to understand issue and how it can be addressed in future as given below-

1) **Use of inferior grade of cables/wires/switchgears/ electrical materials –**

   In India ISI is certifying body for many products. They have formulated testing procedures and specifications for that product. When there is ISI marking on product it means that product satisfies or conforms all the specifications and testing stipulated in that ISI book. Manufacturers have to get their product certified from ISI if they claim that it is ISI marked. In India many electrical materials like switches, sockets, cables, wires and many other equipments are ISI marked. Once marking is done then manufacturer is legally bound to satisfy the conditions always mentioned in that standard.

If ISI standard is not available then IEC standard shall do. ISI has adopted IEC and hence one can opt for IEC also.

Each nation has their own standard. Product manufactured in that nation is tested and conformed as per that standard if product is to be marked.

At the same time, one has to ensure the approval authority certificate is valid.

Standard mentions the specification of materials in which predetermined values and testing procedures are given thoughtfully and with experience.

There are products which do not have the standard marking. Such materials are most likely are substandard. They might be available at lower cost than corresponding ISI mark product but in long run such products are bound to fail.

So, user should remain vigil while using product and insist on usage of ISI marked product only.

At the same time in long run, in practice if the said product is failing then one should avoid to use the product. So, market research about product is also important before usage.

In the competitive world, manufacturers are doing cost cutting in standard product, which in long run affects performance. So, such products although economical are not to be used.

Many organizations select samples out of the lot supplied and verify the product from third party testing agency. This is effective tool and can be adopted.

Sometimes quality guidelines like standard internal process, quality check etc. provided by manufacturers. This has added advantages. So, such products are preferred.

Likewise, electrical contractor assessment is also to be done for their skills, internal quality checks, past record, tools tackles, facilities etc. and such contractors are preferred.

Another commercial factor is of time and money. Quality gets affected due to both these factors. In such scenario one should not compromise for the quality if at all project goes out of hand.

In view of this inferior materials are to be avoided and standard quality good materials are to be used.

Let us take example of wires.

Wire having less insulation – PVC insulation is provided on live conductors in wire so that due to die-electric strength of PVC insulation, live conductors do not get exposed. If live conductor comes in contact with any human being or live creature, then there is danger of shock to them and if live conductor comes in contact with any metal then there is danger of leakage current, spark or indirect shock and ultimately fire. Such is importance of PVC insulation in wires.

Few cases observed are –

1. PVC insulation is not adequate as per standard.

2. PVC insulation thickness is not uniform and less insulation on one side.

3. Dielectric strength of insulation is lower than specified as per standard.

4. Recycled PVC is used.

5. Dielectric strength is inadequate.

If such substandard material is used, what will happen is -

After wear and tear insulation will further deteriorate and then conducting part will be exposed.

Similarly, there are also issues with <u>conductor-</u>

1. In spite of minimum wire size as per standard should be 1.5 sq. mm copper still one come across usage of 1 sq. mm copper wires.

2. Inclination towards flexible conductor instead of stranded conductors.

3. Resistance is observed to be higher than limit.

4. Rerolled copper or commercial copper is used instead of pure electrolytic copper.

By using substandard copper conductor, heating of wire and thereafter insulation burning process will start.

It is advisable to use standard wire makes which are prudent. They will have insulation and also electrical parameters well within permissible limit. The makes used should be again and again checked. The cost should not be the only criteria from safety point of view.

To go in little depth, one has many options to select out of various types of wires. Depending upon properties as mentioned below one can select the best of best wires to make installation robust. Although costly it has advantage of safety. Following are examples of different types of wires.

In wires as per IS 694 and in 17048 standard there are various types like –

- FR having properties of <u>Flame retardant</u> so that the propagation of flame is retarded.

- FRLSH having properties of <u>Flame retardant low smoke low halogen</u> so that there is restriction of the spread of flames in fire situation and the smoke emitted by the burning of cable is considerably low compared to traditional cables. With this there is improved visibility for evacuation of trapped victims and facilitates firefighting operation.

- HRFR having properties of <u>heat resistant flame retardant</u> so that it retards the propagation of flame without compromising safety.

- HFFR <u>Halogen free flame retardant</u> as per IS 17048 having properties such as withstanding temperature. Insulation does not burn, melt and drip. Smoke is negligible, transparent and non-toxic. It is self-extinguishing and flame retardant.

<u>Such variants will be seen in all products.</u> User must select best out of lot. In view of this select material having standard marked, best quality and good track record. <u>Selection points of wires, cables, switchgears are covered in further topic.</u>

Although provisions are made in electrical installation guide lines, standards, CEAR rules, regulation to use <u>standard</u> materials, it is observed that there are failures of materials because of using inferior grade, nonstandard, non-ISI marked electrical equipment or cables or wires and thus electrical installation becomes hazardous. So, it is imperative to use standard quality materials by users. Also, before using, testing of entire installation is required to be done. During testing, defects arising out of manufacturing or handling or while installing can be detected and further damage can be avoided. If one finds any inferior quality / substandard materials then dare to complain to authority without any hesitation.

As a rule, use the material which has marking of the applicable standard, having good track record of the same and good quality material so that chances of failures and likely chances of fire will be avoided.

Summary-

- Use ISI / IEC marked materials.

- Check track record of materials.

- Insist good quality of materials and not only price tag.

- Before using materials carry out testing and if required do through third party agency.

- Overview manufacturer's internal quality check plan and if not in place, then insist for same.

- Check contractor's capabilities such as experience, skilled electricians, tools etc.

- Quality priority should be first than time and money.

- Respect the specification.

- Testing is to be done after installation.

2) **Temporary connection –**

Many a times it is observed that either from street lighting pole or from feeder pillar bus bars provided by electricity board, electrical supply is tapped for temporary work / temporary lighting / festival lighting. Hence loose connection, proper fuses or switchgear is absent. So also, cables/wires are not installed in standard way and lying unattended or placed on temporary structure causing stress on installation. Absence of proper connectors is observed. Following can be seen for example.

Images by Author

There is absence of glanding and no lugs. Sometimes wires are twisted and joints are made. Instead of using Cable jointing kits for joining cables, wires are twisted and joined and tape is applied. Usage of lugs in termination is not being done. At times insulation tape is so temporary that it is coming out or gets washed away. Wires/Cables are exposed to rain, water, heat etc. and protective coverage is absent. All this will cause ultimately unsafe installation and causing sparking and fire.

Summary-

- Temporary works are to be carried as permanent work without compromise.

- Avoid loose wiring / cabling.

- Use lugs glands wherever required.

- Focus on terminations.

- Use adequate size of switchgear.

3) **Improper Construction Power supply –**

For construction sites temporary power is drawn some times as mentioned above. Sometimes it is observed that cabling is so temporary that doors of panels or feeder panels are absent. There is no proper earthing to panel. Earthing pits might not be proper or non-standard. Local damages to cables / wires while installing the same will have bruise, nail piercing, cutting insulation may damage insulation and arise failure. At bigger project sites nowadays, systematic installation is seen and there is lot of awareness among users than earlier. Safety officers are appointed at site who take care of safety parameters on sites. But that may not be true at smaller project sites or at rural sites. Still temporary cabling without switchgears, without proper earthing supply is being used. Non usage of Plug tops, RCD, no usage of proper distribution boards is common. Dewatering pumps without earthing is also seen. Lighting is temporary. Halogen flood lights are used without earthing and improper mounting arrangement is a common scene. This needs to be looked upon by users and they need to remain vigil. In short, handle construction power supply or temporary electrical installation as per same rule as per permanent installation. Else there will be sparking in no time, which might turn in to a fire. Hence it is recommended always to legalize the construction supply or temporary supply from supply company and get the same approved from licensed electrical contractor.

Following photo is from construction site where temporary cabling was done. This is improper installation and ultimately will lead to short circuit.

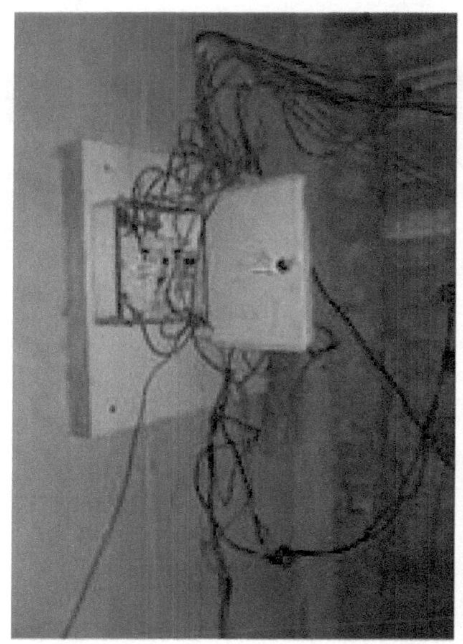

Image by Author

<u>Earthing</u> in electricity is a very important topic. Let there be any type of electrical installation / connection. But when temporary work is done then pit is damaged, continuous earthing is absent, name plates / green color not given for identification. Some of the examples are as given below.

Images by Author

Hence proper earthing as per standard has to be done if at all it is temporary or construction power supply work.

Summary-

- Construction power supply work should be done as permanent work.

- Use RCD, earthing, plug tops etc. without fail.

- Avoid using damaged materials.

- Wires and cables to be laid properly.

- Use adequate size of switchgear.

- Protect switchboard from rain.

- Mounting of switchboard should be permanent.

- Consider views of safety officer.

4) **Select proper Power outlet / Socket capacity –**

Power Outlets / Sockets are commonly used in all applications. Before going in deep, let us understand single line diagram of residence apartment which is quite simple. In residence maximum fire happens at this point. Power is given by electricity board to main switch in meter room. Main switch brings power in premises by cable / wires up to distribution board located in flat. In distribution board there are MCB, RCD etc. for protection, from there wires are provided through conduit and up to socket. And load such as TV, Geyser, AC, Micro-oven, etc. are connected to it. Every outlet has its own capacity as 6Amp, 16Amp, 20 Amp etc. It is expected that load connected to socket should be such that capacity of socket is sufficient to give power to load. In fact, it should be more than load.

When load is connected to power outlet, it will start drawing the power by means of current flowing through it. If load current is more than the outlet capacity, then naturally power outlet will start heating. E.g. if AC is connected to 6 Amp socket instead to 16 Amp then current drawn by AC will be more than socket capacity and associated wiring, means resultant will be wires getting hot or will start burning so also socket. Because there is local heating and it will lead to sparking and fire.

At that time if the protective equipment like MCB located in distribution board cannot isolate the system then excess current drawing procedure does not stop, causing further heating and then spark and burning of contact and or wire will start. This ignition is sufficient to catch fire if combustible material is available nearby.

Arc Fault Detection device AFDD is new product development which is being used in distribution boards for detection of small arcs happening in between same wire because of accidental drilling or due to any other reason in equipment. Same can be also used for AC, Washing machines etc.

As such first select current rating of switch socket as per usage. Cables/wires/switchgear should be as per connected load. Consistently it is observed where contact burning is seen, the switch sockets are overloaded. Work should be carried out by known electrician only and user should not add the equipment as per wish when socket is seen. And electrical contractor / electrician should provide the rating of socket with adequate back up wiring.

Plug tops are another point to be noted by user. It is observed not only by users but at site level or at electrician level that wires are inserted without plug top. This is the major mistake by anyone. Without plug top means loose connection and sparking is bound to happen. Any flammable material nearby means small spark will change to big fire. So, this mistake should not be done as shown below.

Image by Author

Further mistake happens when instead of using three pin, two pins are used. Using two pins means no earthing, it means frame fault cannot be sensed and will result in shock. Such incidences happen in festival lighting mainly. Two issues are happening in this – One issue is, two pin remains loose in three pin socket causing loose connection and sparking. If any flammable material is nearby, which is invariably present in vicinity such as curtain, then likely chances of fire. Second issue is no earthing hence likely chances of electrocution if phase wire insulation burns and metallic part of apparatus becomes live.

At site it is now recommended to use RCD. Even in residential DBs RCD is recommended. So that if there is leakage happening or starts then immediately it will trip the circuit and give adequate protection. In residential complex especially in mass housing, RCD is not used. When RCD is used and there is leakage then it trips. Many users think that tripping of RCD is nuisance tripping, which is not so. RCD is doing its work of tripping if there is leakage current. Users must think that there is fault and so let us find out. Instead of doing this, they will bypass or remove RCD and complain it as bad installation.

Switch socket is most neglected but it is most vulnerable, weakest point in many failures as are seen. Hence if proper care is being taken then fires can be avoided.

Further is extension board issue. If there is one socket and many equipment are to be connected such as TV, Music system, light, PC etc. then extension cord is used or multiple sockets are used on one circuit is used and power is drawn. Same issue as discussed above if load current is more than socket capacity then again heating, burning etc. starts.

Image by Author

Summary

- Select current rating of switch socket as per load usage.

- Select wires as per socket and load rating.

- Do not add equipments on sockets unless advised by electrical knowledgeable person.

- Avoid usage of extension boards.

- Use plug tops and do not insert wires and work.

- Three pin plug is to be used and third pin to be connected to earth.

- Use MCB, RCD in DB as advised by electrical knowledgeable person.

- AFDD to be used for high end loads.

- Any blackening of contact means danger, so attend immediately.

- Joints in wires are not permitted.

- Installation work should be carried out by licensed contractor / electrician.

5) **Under sizing of Cables / wires –**

Cables and wires are selected by design engineers and specifiers as per standard in which it is clearly specified the usage, numbers, temperature and also product catalogue of manufacturers. However, this may not be always true when the same are being used by users. Many a times by experience selection is done. At that point mistake happens.

Let us take example in case of wires. In product catalogue, there are two ratings of wires- one in casing caping and other in concealed wiring. In case of concealed wiring there is more reduction in ampere capacity of wires than that of casing caping. In standard, number of wires in various conduit sizes are given. If electrician is exceeding the number than specified then naturally heat dissipation will be affected and naturally current carrying capacity of wires will reduce.

Let us take examples in case of cables. In product catalogue current carrying capacity is given in three types of installation such as in ground, duct and air. In lower sizes of cables, ampere capacity is higher in ground then it reduces in air and further reduces in shaft. Where as in higher sizes of cables, ampere capacity is higher in air then it reduces in ground and further reduces in shaft. So also, as per standard, there is further reduction in current carrying capacity in air temperature rise, ground temperature rise, also depending up on depth of laying, number of cables in group, number of trays in duct etc.

Apart from above one has to see voltage drop, short circuit calculation, type of insulation, future loading.

Overloading can be seen from following photos. In right side photo, nothing unusual is seen in installation when observed with naked eyes but left side photo under thermal scanner shows that in installation the bus bars are heated and now on verge of short circuit.

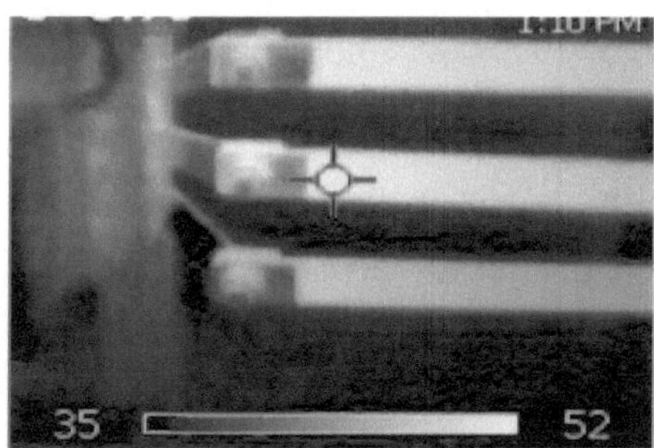

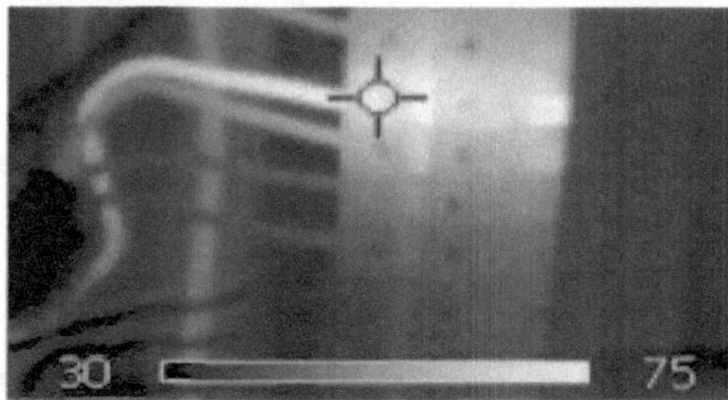
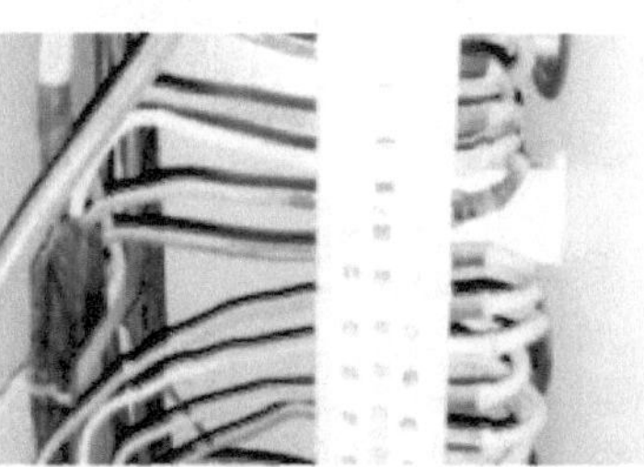

Images by Author

Well, so while selecting the size of cable one has to consider the fact that rating of cables can go to 50% than catalogue rating due to above factors. Hence selection of cable should be proper.

So if selection is done only on basis of catalogue rating, it will result in heating of cables / wires / busbars and then subsequently insulation failure and then further deterioration, electrocution, fire etc. All these issues will erupt with selection of undersized cables. Hence cable, wire sizing is very important.

Summary

- Select cables and wires as per usage mentioned in product catalogue.

- Select rating as per temperature, method of installation, material type, insulation there on etc.

- Consider future loading on cables used.

- Consider voltage drop, short circuit etc. while selecting cables.

- Proper termination is must.

6) **Joints in wiring –**

At times it is observed that in small dwellings, residences, construction sites, exhibitions and even in projects when situation arises that existing wires are short then wires are joined and extended.

Images by Author

This is the most dangerous installation and not recommended to be done instead change wires in full length. To join the wires together, electrician removes part of insulation of two wires near joint and twists the inside conductors together by plier and then ties the insulation tape over the joint. In process there is no guarantee of joint and it can culminate in loose connection. Due to this loose connection while pulling the wires there could be break in wires, which may develop spark or may touch to conduit. This practice is strictly not to be followed at any cost. In unavoidable situation use lugs, ferrules, terminal box or isolator. If size of wires is more then use ring type lugs on both conductors, tie together by nut bolt and apply insulation or alternatively use terminal box and both lugs are fixed on it or if possible, use isolator switch in between. But all this can be done for higher sizes of conductor but in case of small wires like 1.5 sq. mm this is not possible and hence to replace the full wire is the best option but never encourage joints in wire.

In case of cable situation changes because readymade jointing kits are available. But method of joining is to be followed strictly as per kit manufacturer. So, when it is absolute necessary then only use jointing kit. But whenever there is a planned activity, joints should be avoided in project and cable quantity to be ordered as required at site.

Summary

- Do not carry out joints in wiring.

- Any sparking to be avoided.

- In unavoidable circumstances use jointing kit.

- Recommended to change wires instead of giving joint.

7) **Outdated wiring** – One of the main causes of fire is when wiring becomes old and outdated and needs replacement. Reason behind this is deterioration of insulation of wiring over the time.

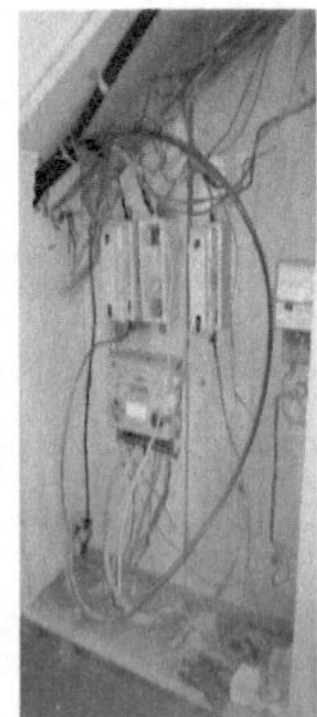
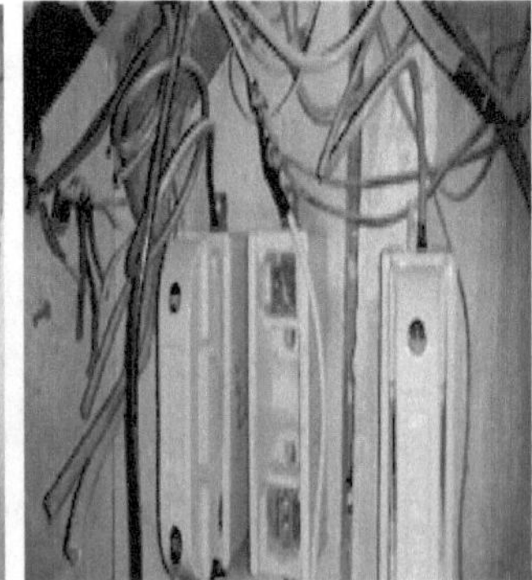

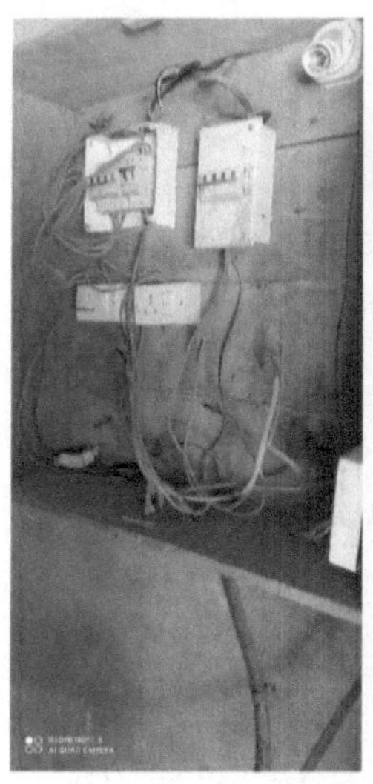

Image by Author

Terminations become loose which can cause short circuit in wires / cables. Loose connection at termination, damaged and or pitting at termination, improper fuse / switchgear ratings, loose bus bars connections, increased load, all these contribute toward hazards.

Following is example of outdated wiring termination and need of its' replacement.

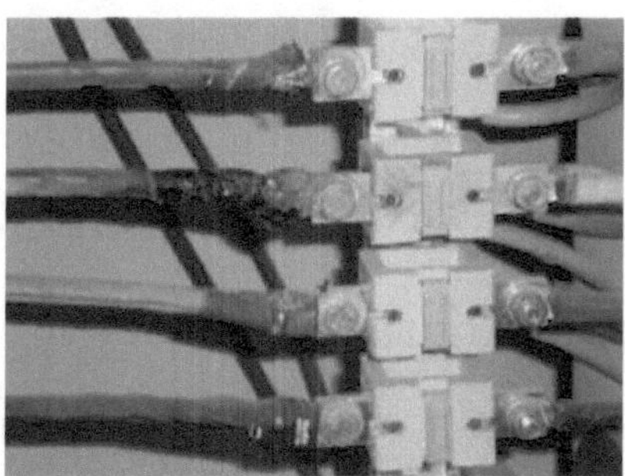

Image by Author

Load has increased over the years because of addition of newer apparatus. As load increases current drawn also gets increased and cannot be catered by old wires due to sizing, insulation weakness and loose connection or there could be old switchgear like fuses, switches and there arise such problems. Reasons are same as above applicable for terminations, operations, life of material and mainly inadequate sizing of cables becomes inadequate load may have increased. At times rating of fuse is so high that it does blow. Sometimes rating is low and when fuse blows it is removed and replaced by copper wires. Copper wires having higher rating do not blow. Both concepts are wrong. In both the cases supply does not get disconnected at proper required time and ultimately load on wires increases and heating starts. If it remains unnoticed then burning starts and in no time, it changes to fire. Wires and cables also have life and one must check the insulation level else will turn into heating.

Summary

- Replace outdated wiring immediately.

- Replace outdated switchgears.

- Replace damaged electrical materials.

- Use adequate capacity of meter.

- Check the load and replace electrical materials.

- Change location of meter room from below staircase.

- Dressing of wiring, cabling to be carried out proper way.

- Carry out ventilation of electrical room.

8) **Residential loose wiring-**

This is the major cause of fire at rural areas or in small dwellings. At times habitually occupants lay wires through beds/carpet/on floor/in air. Temporary wires are also taken from supply company boards or extensions taken from next door residents or rooms. All these things are unlawful and also not standard practices of laying wires. Even sometimes earthing is not provided in such wiring. After some time, there are bound to be wear and tear of insulation of wires. Once insulation deteriorates then there are likely chances of sparking remaining unnoticed or may be undetected happening due to short circuit. This results in electrocution and / or fire.

Likewise, is the case of loose wring. Now a days it is observed that people are keeping laptops that are connected to power socket and wiring is slackly kept on bed / sofa. After the work is over, situation remains unattended. When laptop is in charging mode one can feel the warmth on bed underneath of laptop. In case ventilation is improper or laptop is warming more, in such case sufficient heat is produced for bed to catch fire. Few incidences of fire happened in such case. As such occupants should ensure no loose wires are provided, no equipment which will generate heat should be kept near to any flammable material.

Pump rooms in residential building is part of above examples. Following photo can highlight temporary improper cabling done for pump. Today or tomorrow this loose incoming and outgoing cabling can be reason of short circuit and then fire.

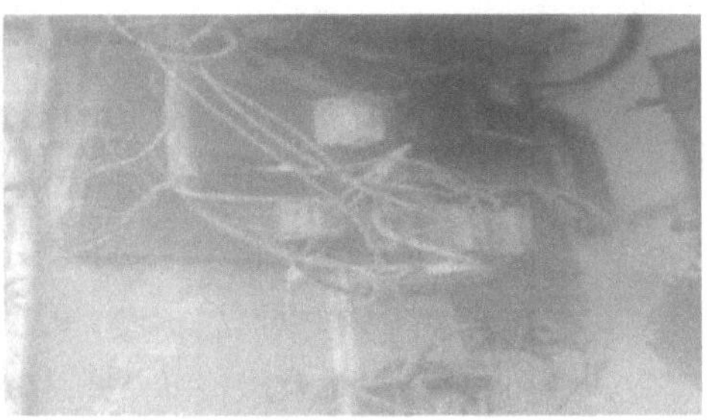

Image by Author

Summary

- Loose wires under bed, carpets etc. to avoid.

- Do not keep laptops on bed for charging

- Avoid extension board.

- Tapping of supply from supply company board is not to be done.

- Instead take regular connection.

- Power tapping from adjacent house is to be avoided.

- Loose wiring in pump room, lift room to be replaced by proper cabling.

9) **Faulty Space heaters/Geysers-**

Space heaters are provided for heating rooms where temperature lowers down in winter. They are needed but nonstandard heaters cause hazards. They generate heat which has no control at times and becomes excessive. If heating coils are old, they can cause internal short circuit and induce sparking. If a fan is there inside the heater, heat radiates faster in the room. Faster radiation of heat may cause deterioration of insulation of wiring in the vicinity of fan. Caution is required here to check whether wiring and insulation is intact and unaffected by heat. Heat and fan speed should be regulated to avid unforeseen incidences. Similarly, in case of nonstandard geysers if there is no protecting valve or cut off mechanism then chances of explosion are there. Sometimes after repair of geysers wiring is not properly done and may cause internal short circuit. As such gadgets are required but gadgets of inferior quality and without safety measures can also cause explosion or fire.

Summary

- Use reputed brand ISI marked heaters, geysers.

- Check wires, fan, coils if proper.

- Check device for cutting off power supply, safety valve is provided.

- Keep away heaters from bed, curtains or any flammable materials.

10) **Faulty AC –**

Figures of fires due to air conditioners are now a days swelling. There are mainly three types of issues in fires due to air-conditioning, and they are electrical, mechanical and maintenance. Wires laid for machines can get old or damaged due to brushing of wiring on mechanical edges of machines or window etc. This results in worn out insulation that may turn into spark and short circuit. In some cases, extension cords are used and if they are not of adequate capacity, they create heating and short circuit. Other than electrical issues, there are mechanical issues such as motor bearing is not proper and hence motor over heating or burning, wobbling of blower causing spark. Or any other problems that lead to overheating. If maintenance is not done regularly and timely then

AC filter starts clogging and users start lowering the temperature parameter for getting result. So, motor starts getting overloaded and therefore may burn out. As air conditioner gets older its' motor derates and wears out and this can be source of fire. Hence better to replace old machine before it fails completely.

Also, If the machine is not cleaned inside, and if flammable materials like dried leaves etc. are there inside, they will burn. Debris, leaves, dirt etc. inside the machine and over the motor can insulate the motor or their presence inside the motor leads to friction and subsequent overheating. If flammable materials such as dry papers are on the outside of the air conditioner, it creates a risk of fire hazard as one small spark coming out of machine will ignite the papers.

Fires due to air-condition machines is worldwide a worrying factor because as per NFPA, it was found that yearly 2,800 air conditioner fires take place in US only with property loss of about $78 million per year. And in Delhi nearly 15% fires are due to ignition of air-conditioners.

In recent pandemic there were shortage of hospital beds for patient. Due to which it was necessary to open up temporary facilities/centers for patients. AC machines put in service were running 24x7 without any maintenance or redundancy causing these machines get overloaded. Existing AC were used in hospitals where beds were added and due to which heat load has increased and AC machines were working without any stoppage. In critical areas like Operation theaters, ICU, laboratory etc. AC machines which are installed are working round the clock without any maintenance and there is no provision of redundant machines. In such case these machines are bound to fail or start developing mechanical wear and tear. Another issue was electrical power consumption in rooms were increasing because of addition of medical equipments, additional temporary AC etc. but main line or switchgears or cabling were not enhanced to higher capacities causing overloading and overheating of cables which prone to sparking and fire. Due to added oxygen cylinder to each patient oxygen content in room got increased and due to disinfectant air content became flammable. Hence hospital environment was susceptible for fire. Hence AC machines needs maintenance and should be put off at frequency as suggested by manufacturer.

Sample piece of indoor part of AC after burning is shown below. This burning of AC will lead to a huge fire.

Image by Author

Summary

- If wiring of AC machine is damaged then replace it.

- If plug top is damaged then change it.

- Maintain AC machine as per manufacturer's guidelines.

- Clean filter regularly.

- Give rest to ac machine as per guidelines of manufacturer.

- Change parts like fan blade, motor, coil if they are damaged or worn out.

- Use proper sizing of ac machines and do not overburden it.

- Ensure ventilation for outdoor machine.

11) **Placing paper/cloth on lamp shade** –

It is observed that in open markets or in small homes people frequently use papers over the lamps instead of using lamp shades. Such papers can get ignited due to temperature of the lamp. One may find this reason frivolous, but many fires have broken out due to this reason. Majority of the incidences have occurred earlier where GLS, Halogen, vapor lamps were present. Burning papers or cloth can cause further deterioration and chances of fires increase due to short circuit by loose wiring, heating of insulation, not adequate protection. Now a days because of LED usage, such probabilities are reducing in urban area but in rural areas still chances prevail and hence care needs to be taken.

Summary

- Use proper lampshade.

- If you observe vendor using makeshift shade by paper then inform to remove.

- Select tested good quality of decorative light having fabric or plastic cover.

12) **Issues with Meter / Electrical Room and shafts –**

In past many fires broke out in meter room and in shaft. Shaft acts as chimney and spreads fire throughout the apartment/building. Reason being old wiring / old switchgears in meter room, old busbars which are under capacity, broken insulators, heating of busbars due to loose terminations, area being insufficient and hence busbars switchgears are getting congested, no ventilation. Place of meter room under staircase / under toilet, unclean meter room, no maintenance of electrical work, copper wires used instead of fuses, proper earthing not maintained, loose terminations, insufficient capacity of meters / fuses / cables etc. which are not augmented as per increased load capacities are also major reasons. Shafts are not closed on each floor and are not separated, cables / wires dressing is not done, congested installation, increase in temperature, Installations and materials of other services also present in   electrical shaft, water seepage in the meter room are also common situations observed.

Sample photos of bad meter room is given below-

Congested Room

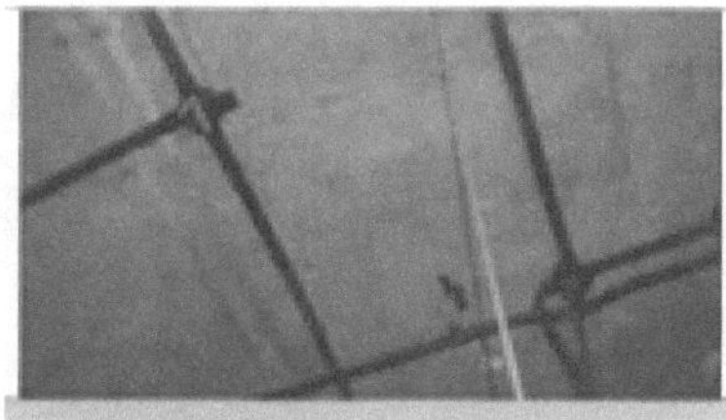

Openings in conduit junction boxes, boards are kept open.

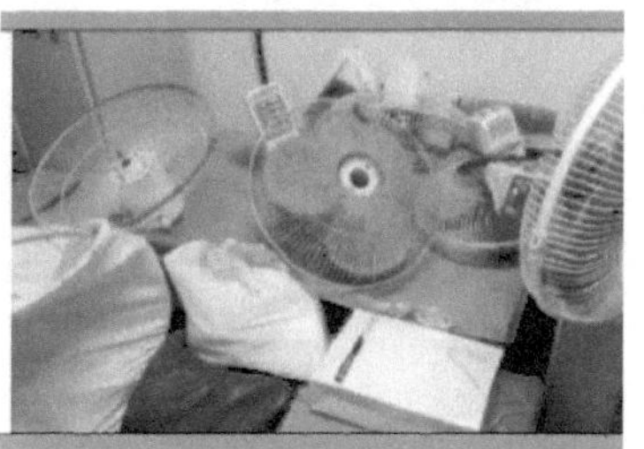

Electrical Room having panel is dumped with materials used as store.

Substation is full with tree, weeds etc.

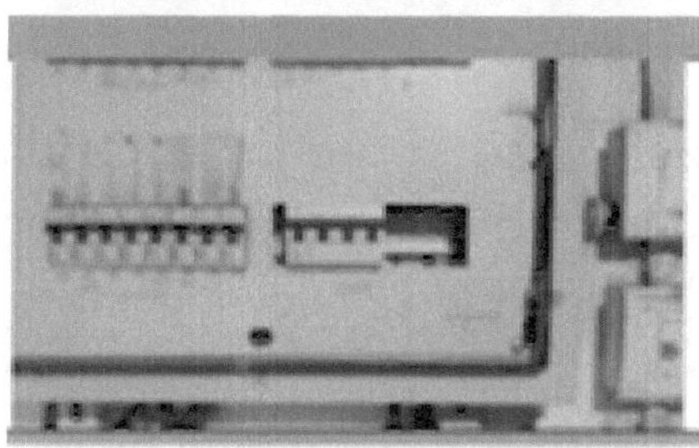

Distribution boards are open and not closed.

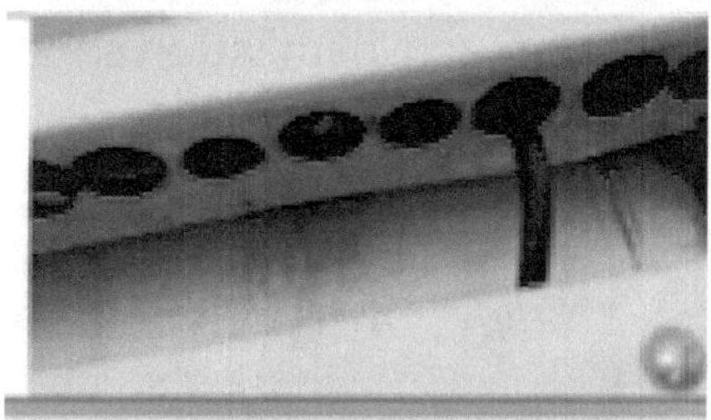

Opening of holes causeed dust to go in or becomes lizard fault as per right side photo.

Cabling without gland, improper way of cabling.

Images by Author

Electrical shafts are one of root cause of fire in residential building because of congested shaft, improper cabling, dressing, and clamping of cabling not done. Following photos can throw light on the same topic. Such type of installation will make cable current rating go down and so load heating will start and then short circuit. Hence such installation should not be done.

Images by Author

Above points and effect of the same were already discussed in details. Chances of sparking, heating and therefore risk of fire breaking out increase because of all these issues. Serious thought has to be given for meter room, shaft, and installation there in to prevent fire.

Summary

- Replace old wiring, switchgears and busbars.

- Check the load and augment the capacities.

- If heating is happening then inform electrician and ask to attend immediately.

- If there is congestion in meter room then remove it.

- Keep electrical meter room, shaft ventilated.

- Remove meter room under staircase to proper place.

- Use proper sizes of fuse and not wires.

- Close the shaft by fire sealant after laying cables, busbars.

- Not to use electrical room as store room.

- Plug the holes in panels, boards.

13) **Temporary/Mobile DG sets –**

Fires have taken place due to temporary DG Set. There are good number of reasons why it can happen. Some of the main contributors are worn out parts, reduction in insulating resistance, earthing and poor maintenance. When DG is provided near pandal or at hall, permanent earth pit is not made. So, what is being done is dig and burry GI pipe rod claiming it to be earthing rod pit or alternatively to connect to nearby electricity board feeder pillar earthing. In fact, there should be two distinct earthing for star point and two for body. But this rule is compromised. If earthing is not proper and high resistance earthing path is present then chances of tripping of the switchgear in case of fault are rare. Consequences of not tripping of switchgear will be heating of the cables in case of cable fault or heating of alternator in case of alternator fault or heating of any other circuitry which has fault. There can be chances of electrocution in case body / casing has leakage current. It may lead to fire also because of presence of flammable fuel such as diesel if present nearby. Many DG sets are not overhauled properly and hence changing engine oil, coolant, belt replacement etc. are pending.

Photo indicates the set which is not maintained, cleaned and pressed into service. So, it is going to fail and may even burn.

Images by Author

Result is burden on engine and subsequent overheating of the engine. As temperature increases, any loose wiring termination if creates spark then it turns into major accident because of flammable diesel present. If there is leakage of diesel then there is chance of fire in case of spark or heat and also because of its flammability. DG Set should be provided with control panels as per standard. However, in practice it is rarely provided. Generally loose copper unarmoured cables are taken to mains supply by laying the cables on floor. Absence of permanent change over switch at mains, calls for another loose connection. Also cabling becomes loose without proper glanding. With this, chance of sparking and short circuit increases. Due to absence of proper foundation, vibration of DG set is transferred on van which is carrying DG Set in canopy. Due to this arrangement vibration can loosen the connection and hence maintenance of set is required. If not done then installation is not healthy.

Summary

- Installation of Mobile or temporary DG is same as permanent installation.

- Replace worn out mechanical parts like fan belts, fan, guard.

- Ensure sufficient coolant and good quality of oil.

- Avoid vibration by antivibration pads.

- Arrange proper earthing.

- Loose cables to lay and protect properly.

- Check the terminations.

- Attend the leakages immediately.

- Use well maintained Set.

- Periodic maintenance and checks to be carried out.

- Duty hours to be maintained as per guidelines of manufacturer.

14) **Over loaded transformer** –

Many fires broke out at transformer location. Transformers should be loaded within their capacity and not to be overloaded. At times Electricity boards are under stress to release power especially in rural areas and uncontrolled load comes on transformer which causes blowing of DO fuses. If proper load management is not done or higher size fuses are installed then starts overheating of transformers followed by winding failure and finally explosion takes place. Many transformers are not over hauled in time and filtration is not carried out. In absence of filtration oil is not clean, slurry is formed and dielectric strength reduces. This results in lowering of flash point and possibility of overheating oil and then burning. There is no early alarm system like Buchholz relay for smaller transformers. Therefore, if there is interturn fault then there will be no alarm or no tripping except DO fuse protection for higher current. So, what is important is the maintenance of transformer. In power transformer maintenance and over hauling is very crucial from the point of sparking, failure and fire. Corona effect means bluish glow is seen due to breakdown of air around high voltage at bushing terminals of transformer and needs to be attended and verified else sparking of the same will create further damage. Silica gel has to be replaced timely else moisture will start entering in oil causing further damage. Oil filtration is by default is a must but checking pH level / sludge are also important factors from dielectric point of view. If there is sufficient ventilation or no is to be checked and for that matter proper working of the fans also needs to be checked. All safety relays and protecting devices need to be verified which are giving protection to transformer from high temperature, internal faults. They give guidelines and pointers towards faults and same have to be respected and attended immediately. If any servicing or unscheduled maintenance is required then it has to be carried out without postponing else small fault can lead to disaster. Terminations of power and control jointing need to be tested as per maintenance schedule. Dusty atmosphere reduces creepage distance resulting in short circuit. Hence cleaning of bushing is required at sites wherever dusty atmosphere is predominant. Insulation resistance of windings is also important to be checked because after years of usage insulation level of winding deteriorates. Hence if timely checking is not done then inter turn fault occurs and results in internal short circuit. Timely checking for turn ratio, insulation resistance, oil sample testing as prescribed are required. Maintenance prior before monsoon is required to be done otherwise faults will be invited inadvertently which may further aggravate to possibility of short circuit. Protection against lightning is to be verified. Earthing grid values are to be checked as per schedule. Power transformers are protected by means of nitrogen injection fire protection system. In spite of that fire incidences happen. Reason being not well-maintained transformer and protection system. As such load management, capacity, transformer oil filtration, adequate protective switchgears, transformer maintenance is must else it becomes a fire threat.

Summary

- Check loading of transformer

- In case of DO fuses replace with correct size.

- In case of switchgears check settings of releases as per design.

- Check insulation resistance.

- Check always dielectric strength of oil.

- In rural areas ensure transformers are not covered by trees.

- In case of large size transformer silica gels to be replaced timely.

- Attend the alarms immediately and take corrective actions.

- In case of high temperature check the transformer without delay.

- Attend leakages immediately.

- Bushing cleanliness and lightning protection to be done.

- Check timely terminations.

- Timely overhauling, maintenance to be done.

15) **Improper termination-**

In many small or big electrical fires majority incidences originate from terminations. Termination are at small to big levels like at switch, socket, switchgear, busbars, battery, wires, cables etc. Improper termination of wires/cables, not done as per method prescribed can be main issue of sparking. There are two main components in termination. One is materials used in and secondly prescribed method. In smaller terminations like in case of wires main issue happens that lugs are not used and directly wires are inserted in to terminals of switches sockets. While doing so at times all conductors of wires are not being inserted in socket provided or because of flexible conductors while tightening small screw few conductors pass out of screw base, making loose connection. Instead, if lugs are used and then tightening is done chances of loose connection are never there. When there are loose connection then it gives rise to local heating, sparking and ignition, which is starting point of fire. In case of larger terminations like in switchgears or in busbars then there should be usage of glands, lugs, inhibiting compound, washer, spring washer, insulation tape, bimetallic washers and importantly skilled manpower.

Image by Author

After joint is complete then tightening and torque as specified to be applied and marking is to be done on it to know that said joint is done by skilled person and checking is done.

One can observe in following photo spring washer not used which is not right way of doing termination.

Image by Author

In case of busbars same process is to be adopted else if inside panel joints remain loose then local heating starts. This is mainly applicable for fluctuating load where heating and cooling effect makes joint weak, if termination is properly not done. There should be no short cuts for this. Many fires break out due to improper termination because it causes loose connection and once higher current starts flowing, larger spark flushes and within no time it turns in short circuit and fire.

Photo below shows the improper termination at transformer-

Image by Author

Few examples are given below in which there were loose connections Same could not be seen by naked eyes but in thermal scanning camera they were visible. Once it is recognized, corrective action of tightening or changing can be taken and fire can be avoided.

Right side shows actual photo seen by naked eyes and left side photo through camera.

Temperature increases due to loose connection from ambient temperature to 65 degree C and higher which is the starting point of insulation failure, burning and starting of fire. Right time action is required in such incidences.

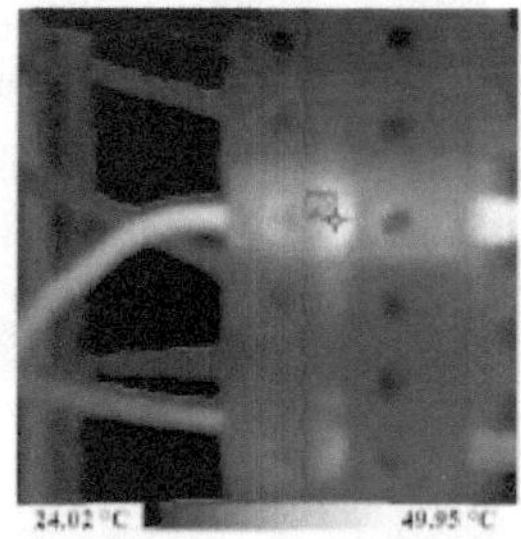

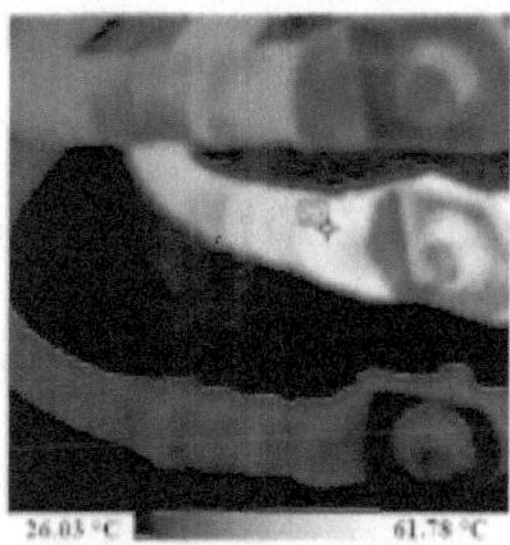

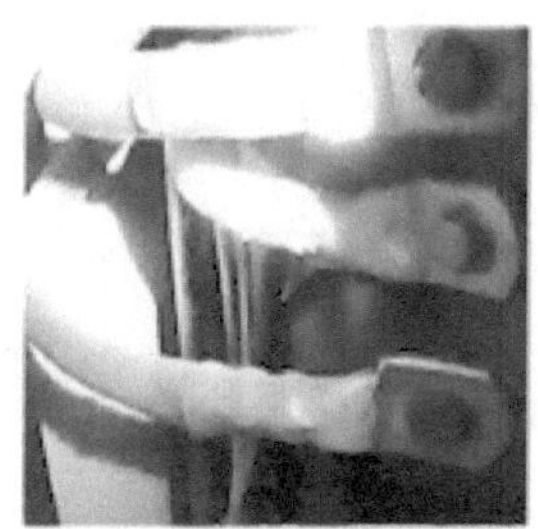

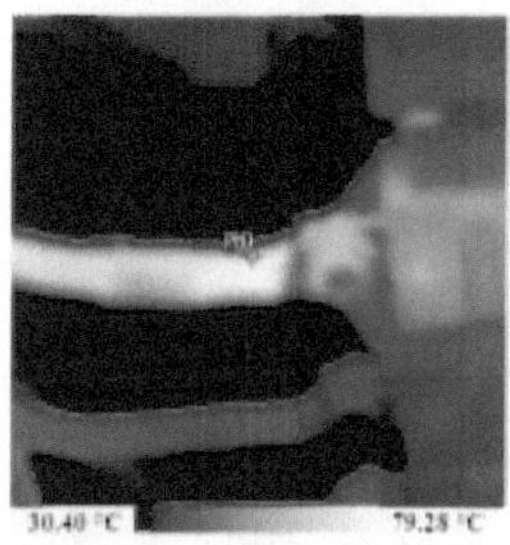

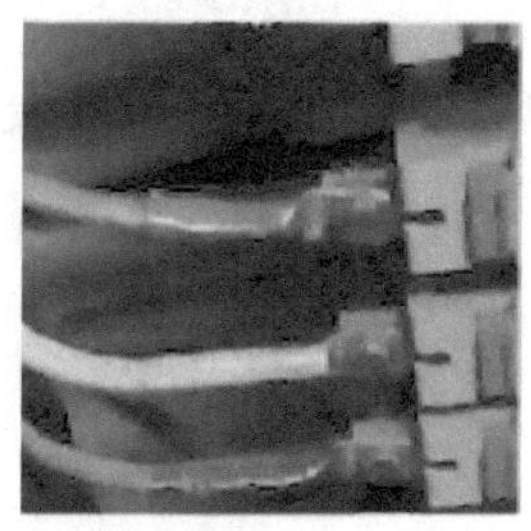

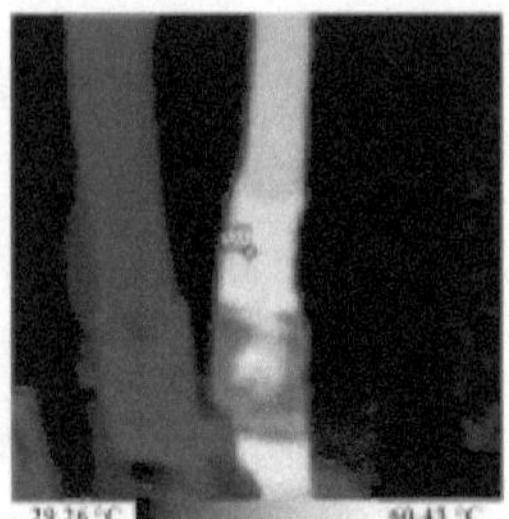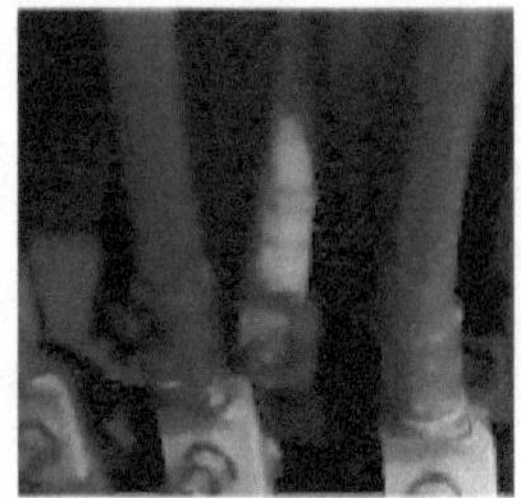

Images by Author

Summary

- Carry out termination as per prescribed method.

- Use compatible size, type and material of lug as per switchgear.

- Use inhibiting compound, washer, spring washer, bimetallic washers as required.

- Ensure all conductors are fitted in to lugs.

- Final tightening to be done by torque wrench.

- Ensure area of contact of lug and busbar.

- Attend loose connection immediately.

- Put off load and attend immediately in case of warm or heated panel, boards etc.

- Use thermal scanning camera to check terminations once in year.

- Remember always termination is very important in entire power circuit.

16) **Static Electricity-**

All substances are combination of atoms. Atoms include nucleus made up of electrons and protons. Normally number of negative pole electrons and number of positive pole protons are same. During the friction, electrons of one substance jumps to other substance. So, substance from which electrons has jumped becomes positive pole charged and substance on which electron has transferred becomes negative pole. Due to this difference static electricity is generated. Due to this spark generates.

Unless static electricity is safely discharged or transferred to another object it remains on a surface. While finding path to another surface, spark is created. It mostly happens with a person walking across a surface or touching another person / metal object. Also, it is created by materials in motion, such as the movement of fluids. When liquids move through pipelines or hoses, the friction creates static electricity. Filtering, stirring, pouring and pumping liquids creates static electricity. When flammable liquids are moving and static electricity is created then it can cause fires and explosions. Combustible dusts can also create static electricity which can further produce spark to ignite combustible dusts. Similarly, operations such as conveyor belts, fan, pulleys, motorized vehicles and bulk chemical storage create static electricity. If humidity is less then static electricity is generated because moisture in air is good conductor of electricity which absorbs excess static charges.

To overcome static electricity in industry worktables and equipment are to be connected to ground, to cover work tables with unpainted metal plate / conductive mat which is to be connected to ground, use conductive floor or spread a conductive sheet on the floor connected to ground.

Operators can use wrist bands, wear antistatic work clothes and conductive shoes which are to be grounded. Use humidifier in workplace if humidity level is less. Also, antistatic shoes are available which are used.

When two objects are bonded together chances of creation of spark while jumping, between two objects do not arise since both are at equal potential. Hence pump is bonded to drum having flammable liquid before pumping the flammable liquid into it and pump/drums are properly grounded. This is applicable in offices or data centers' false floor also. In this False floor used is to be earthed properly so that accumulated static charges can be discharged else likely chances of sparking. As such any static electricity generated must be discharged to earth.

In following exhibit it is seen that false floor posts are not earthed.

Image by Author

For handling static electricity in non-conductive materials humidity is increased. In factories / petrol pump permanent earthing rod arrangement is done for safe unloading of fuel to avoid accidents. So utmost care has to be taken against static electricity to avoid fire.

Summary

- Static electricity can be created through movement of materials or fluid.

- It can be also created through stirring, pumping, filtering, pouring etc.

- Arrangement to discharge static electricity to be done in consultation with experts.

- Use earthing, conducting mat as required.

- Use antistatic wrist bands, antistatic work clothes, antistatic shoes in factory.

- Keep humidifier as required.

- False floor in server room to be earthed.

17) **Harmonics** –

In Linear loads with three phase neutral distribution and at balanced load condition neutral current is almost zero. In practice neutral current is present up to less than five to ten percentage of phase current in case of unbalanced phase condition. But neutral current is never more than phase current. Hence in practice neutral is taken as half size of phase current in case of three phase neutral distribution. In single-phase distribution, neutral is of same size of phase current.

But the scenario changes when connected load is due to having computers, ups, vfd, vrv, servers or any other smps load etc. All these types of loads generate harmonics which are nothing but voltage or current at multiple of the fundamental frequency of system. Most dangerous harmonics out of all is third harmonics which are generated due to computers, smps supply. In case of presence of third harmonics neutral current is almost double than phase current. So, wherever load is such which can generate third harmonics then switchgears, cables are to be selected properly. Because generally neutral sizing is half or of same size of phase current. In such condition neutral is bound to burn since it will have more current than it can carry. One can see from following example where

Phase current is 30A/28A/25A and neutral current is 47A which is almost double than phase current. In this case if 3.5 core cable is selected on basis of phase current then neutral core of cable is bound to burn.

Name	AVG	MIN	MAX	Units
A1 rms	30.265	25.500	32.000	A
A2 rms	28.827	25.500	29.500	A
A3 rms	25.874	20.500	27.000	A
AN rms	47.018	45.800	47.900	A

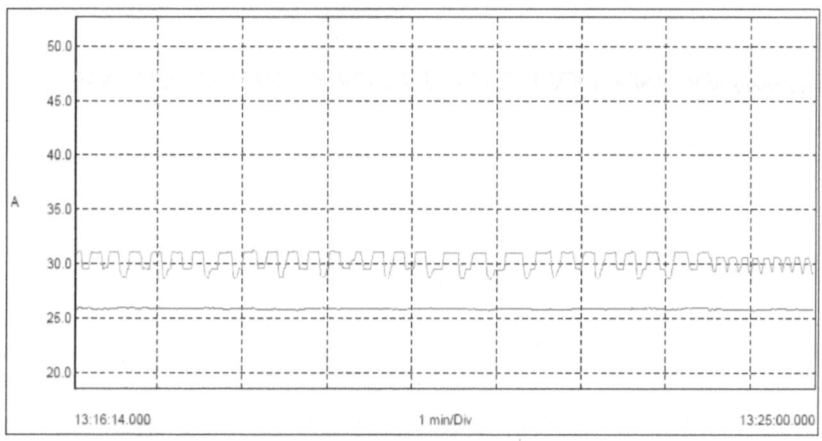

Photo from Powermatrix

Similar is the case of switchgear. At such condition neutral overload / burning cases arise and causes local heating and burning of cable, switchgear and then turning into fire. Hence in case where harmonics are predominant, sizing has to be done carefully for cables, switchgears, busbars accounting for double current in neutral than phase.

Summary

- Use higher size of neutral size of cable, switchgear, busbars wherever nonlinear load is present.

- Remember neutral can be double the size as phase.

- Check transformer end neutral size as per load.

- Use harmonic mitigation techniques to control harmonics.

- Carryout power audit and find out the harmonics and take actions through experts.

18) **Lightning –**

Lightning is spark discharge or short-lived electric arc happening during thunderstorm, in which spark discharges take place from one cloud to another or between a cloud and the ground. Lightning strikes are of a magnitude between 2 and 200 kAmp. Lightning is most dangerous phenomena and it discharges high current and produces high potential gradient dangerous to persons and animals. It momentarily raises the potential of the protective system to a high value. Nearly 42,500 people were killed due to lightning strikes between 2001 and 2018, according to NCRB data on accidental deaths in India.

When lightning hits buildings, trees, or even the ground itself, the lightning current enters the ground, and voltage at that point increases and voltage reduces as distance increases. As such there is potential difference. This difference in potential puts people and animals at risk of electric shock. Apart from direct strike there are surges which occur up to two kilo meters due to lightning strike or even due to switching operations in large electrical systems. Due to these surge voltages, electronic components get damaged permanently.

Lightning / surges can cause burning of wires if adequate protection is not there. Protective system may rise to high value than adjacent metal causing flashover, lightning current discharges through internal wiring causing damage to wiring - components and may also cause risk to structure, occupants. If cross section of protective conductor is not used properly than thermal effect causes on conductor. The shock waves can cause harm to structure. In view of this Air termination rod, down conductors, joints and bonds, testing joints, earth terminations and earthing should be properly designed and executed as per standard. Since along with direct strike there are indirect travelling waves through cables, lines and hence surge protective devices at entry point in panels, distribution boards to be provided for protection. In past many fire incidences took place by lightning strike. There were electronic equipment damages, and loss of lives due to lightening. In short, proper lightning protection system is to be provided else short circuit is surely expected and further fire.

Following are few examples of such damages –

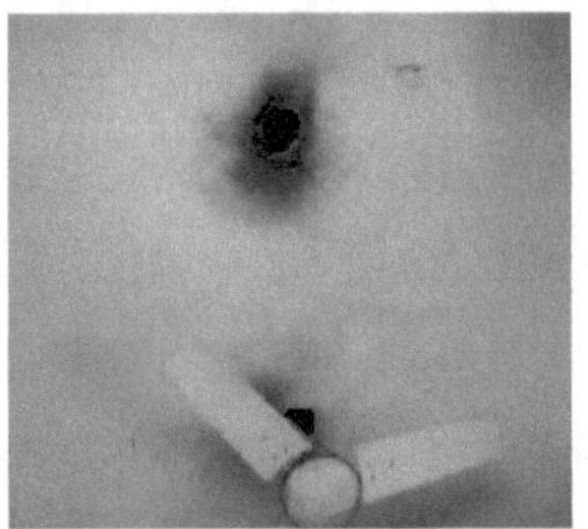

Damages due to lightning, surges.

Summary

- Find out risk of lightning through experts.

- In case of risk is present then design and execute air termination rod, down conductors, earthing as per standard with the help of experts.

- Use surge protective devises in panels and in boards.

- Connect all metallic conductors to protective conductor.

- Adopt equipotential bonding system for earthing.

- For high rise building there are different rules and hence follow the same.

19) **Capacitors** –

There are many incidences of failure of capacitors resulting explosion and fires. As per photos given below-

Images by Author

Reasons being - protective fuse blows due to short circuit inside a capacitor unit, there is overcurrent due to an overvoltage or increase in harmonics. If it remains unnoticed during capacitor failure that capacitor can is bulging or rupturing, then it blows off subsequently. Excessive temperature due to ambient temperature or due to other adjacent equipment or lack of ventilation can also make capacitor ruptures. Resonance happening due to inductance of transformer or due to harmonics in circuit with capacitor and inductor then capacitor is stressed and cause failure. In series connection of capacitors if one of the units fails then voltage on balance capacitors increases causing further failures. There could be manufacturing defect which may cause failures. Capacitor current needs to be checked periodically and as soon as there is reduction then it should be changed.

Ultimately explosion of capacitor is bad event and causes fire.

Summary

- Select voltage level of capacitors.

- Use right capacity, type and standard specification capacitors.

- Check if overvoltage and harmonics in circuit are present.

- Bulging of capacitors or blowing of fuses are early warnings of failures.

- Compartment should be well ventilated.

- Ensure resonance is not happening by using or available inductance in circuit.

- Check always capacitor current and note the same.

- If capacitor current reduces then investigate and if required replace.

20) **Battery-**

Battery usage is increasing rapidly due to rise in telecom sector, data centers, IT, banking sectors. Also, it is widely used in industrial, residential inverter, electronics applications, marine, generators etc. Batteries used are lead acid type, sealed maintenance free, nickel cadmium and now lithium ion. Irrespective of type and usage of batteries the issues related to batteries and accidents, fires associated with them are more or less same. Batteries are in groups and in high capacities and hence issue aggravates very fast and damage is multi-dimensional. If one battery explodes from bank of batteries then like firecracker entire bank explodes making major fire issue. Reasons of failures are discussed here so that care can be taken and catastrophic issue can be avoided. Open Battery Terminals is one of the major reasons although looks trivial. In past fire due to open Battery terminals have happened because of negligence. Since they are not protected by cap, any accidental falling of tool on terminal caused shorting and created the spark which is good enough for further explosion and chain of explosion of bank and big fire. Over charging of batteries is also dangerous. One must be careful while putting Batteries on boost charge when they arrive to site from factory. If batteries are overcharged then they are prone to bursting. So, cut off mechanism must be in place so that as soon as batteries are charged, they should go in trickle mode. Battery room cleanliness, ventilation, dust free atmosphere, presence of AC wherever required, safety equipment, hydrogen sensor etc. are to be provided and then only battery casing / packing should be opened for installation.

Damaged battery can be seen-

Image by Author

Following photos will show that battery terminals are open and not protected and no cover to rack. This should be avoided.

Image by Author

Housekeeping if properly not done, presence of dust, improper ventilation, not maintaining room temperature cause battery to function abnormally. Imagine dust is present in battery can one clean battery top, not possible. It remains in days and likely short circuit takes place. See the picture below-

Batteries are seen but at the same time wooden cartons are still present. False ceiling work yet to complete. Hence room to be completed in all respect then only batteries to be commissioned.

UPS and batteries room not kept clean.

Images by Author

In case room temperature increases beyond battery operating temperature while charging batteries, there will be surely battery failure. If there is leakage of hydrogen and ventilation is absent, Hydrogen remains in room. In such case, one spark and entire room will explode. All reasons mentioned should be handled very carefully and good practices should be followed strictly else these turn out to be sources of fire.

Summary

- Do not keep battery terminals open and use caps provided by suppliers.

- Never overcharged the batteries.

- Keep Battery room clean, ventilated and dust free atmosphere.

- Wherever AC is required, provide the same in battery room.

- In battery room do not keep materials other than batteries.

- Cover the battery racks from top by nonflammable MS cover.

- Bulging of batteries are early warnings of failures.

- Use load banks to check battery cell voltage before putting in to service.

- Wherever possible use battery monitory system to check healthiness of cell.

- Provide hydrogen sensor and fire suppression in battery room.

21) **Faulty wiring on hoarding-**

Hoarding word is used in two ways. One is for advertisement hoarding and another is for things kept cluttered such as newspapers, magazines, personal papers, clothing, appliances etc. Both have increased risk of fire due to short circuit and combustible materials. Any small spark can turn into big fire due to hoarding hence clutter is to be removed and disposed off. Faulty installation can cause radiated heat or spark which can propagate fire if combustible material is nearby. As such safe distance to be kept or if possible, in separate room. In case of flammable materials, flame proof equipment is to be used. In past incidences happened of electrocution where hoarding installed on bus stop got short circuited and bus stop railing got live and electrocuted person. In another incidence banner was mounted on flexible cable and short circuit happened and fire took place. Glow sign box in which tubes are installed is another major issue and should be properly protected else will call for accident. Although there are rules to provide good cabling, switchgear etc. for hoarding, any compromise of quality or installing of the item origins for fire. Especially cables selection should be of higher capacity than requirement because cables are exposed to all seasons and sun light, so deterioration of insulation is worse. Hoarding on bus stops or on buildings if improperly done then inside wiring or lamp source can emit heat or spark and will endanger life of human beings or culminate in fire.

Hoadings which are backlit by FTL tubes are still in use as per photo given below increasing fire and electrocution risk. Because wires are loose, tubes are loosely mounted, backside is steel plate which can be charged any time because of loose wiring etc.-

Hanging tubes and loose wiring on metal board

Cables directly going from footpath without protection and laid under hoarding.

Images by Author

Summary

- Remove and avoid the clutter.

- Isolate the power source from metallic surface of hoarding.

- Use permanent wiring for lamp source in hoarding.

- Earth the board properly.

- Install power supply cables for hoarding in proper way as designed.

- Provide the RCD wherever possible.

22) **Non flame proof wiring in flame proof area –**

Areas in which flammable gases, vapors, flammable liquids, combustible dusts, ignitable fibers etc. are available those areas can be said as hazardous locations. These can be industry, building or part of it. While selecting electrical equipment one has to understand thoroughly hazardous location at which same will be used and location where there are possibilities of fire or explosion hazards. Hazardous areas are classified in zones based upon the frequency of the appearance and the duration of an explosive gas atmosphere. They are then classified as Zone zero, one and two. So also, there are explosive limits such as lower explosive limit and upper explosive limit. Wherever flammable liquid is present the areas are divided in classes as Class A, B and C. Accordingly flame proof equipment are to be used as per types given in standard such as d, e, m, n, o, q with temperature class of equipment T1 to T6. This equipment can be said as intrinsic Safety 'i' in which there will be restriction of electrical energy within equipment and interconnecting wiring exposed to an explosive atmosphere will be having level below that can cause ignition by either sparking or heating effect. On basis of equipment are selected for group IIA, IIB, IIC.

It is suggested that electrical equipment should be located in non-hazardous areas. Where it is not possible to do this, it should be in a location requiring the lowest equipment protection level. As a rule, the neutral and earthed conductor shall not be connected together. Protection against detection of overload, short circuit, earth fault shall be provided. Temporary bonding by earth connections for control of static electricity and potential equalization is to be carried out. All exposed conductive parts shall be connected to the equipotential bonding system covering protective conductors, metal conduits, metal cable sheaths, steel wire armoring, glanding, lightning protective conductor, metallic parts of structures etc. Openings should be closed by fire sealants. Emergency switch off outside of hazardous area is to be provided.

In short designated flame proof areas should be honored and wiring or electrical installation should be flame proof as per required gas group and other requirements. Otherwise Small mistake will turn in to big issue of fire.

Summary

- Identify flame proof area by zones, temperature, group and class.

- Provide the electrical equipment as per the designated area.

- Protection against overload, short circuit, earth fault to be provided.

- Bonding to be done for all exposed conductive parts.

- Adopt equipotential bonding system.

- Installation to be carried out strictly as per flameproof installation guidelines.

23) **Welding** –

Fire incidences during welding are triggered by sparks, hot slag (droplets of melted metal) and torch flames. According to the American Welding Society welding sparks can travel up to 35 ft. (10 m) horizontally and even farther when falling. Spark can pass through or stuck in cracks, clothing, pipe holes, and other small openings. Generally, at 35 ft. sparks can be hotter than 2,500 degrees F. Similarly torch flames can ignite substances within several feet of the flame. As such before welding one has to check for presence of nonflammable / combustible materials and surrounding areas must be kept clean. Generally, at welding site cleanliness is to be maintained and materials like cables, wires, rags, cardboard boxes, paper bags, wooden dust, gas cylinders, wood, and cans of paint, solvents, and cleaning products etc. at site which are combustible are to be removed. See below photo in which necessary precaution not taken and may lead to fire. Such incidences should be avoided.

Images by Author

Openings in equipments, shafts, panels etc. are to be closed else falling spark can travel through opening and creating more damage at other place. And importantly fire extinguisher should be available nearby, welder must wear protection kit and safety officer should be present to ensure all precautions are taken. Main cable and switchgear sizing should be adequate for welding machine as recommended by supplier and importantly earthing wire sizing should be adequate. Fire incidences have happened due to improper earthing or loose earthing connection. Sometimes temporary earthing which is essential for welding is done by attaching piece which is to be welded to large steel & this may cause sparking. Hence proper way and sizes of earthing are to be used.

In past it was experienced that welding can cause fire as such proper precaution needs to be taken.

Summary

- When welding to be carried out keep fire extinguisher nearby.

- Ensure all holes, openings are closed as welding droplets can go down and lead to fire.

- Remove flammable materials from vicinity of welding area.

- Main cable, switchgear including earthing size of machine should be adequate.

- Check insulation level of welding machine before use.

24) RODENT

It is observed at many places rats chew the wires and removes the insulation. They have sharp teeth and grow continuously. Hence they go and try to hide in small places. In such locations if they find wires then they start grinding the teeth. Since wires are round and hence it is easier for them to hold and grind. Eventually they chew the insulation and then conductor. With which conductors becomes exposed susceptible for either leakage or short circuit between conductor and ground or between conductor. If it remains unnoticed and volume is big then spark and fire can take place. So devices like electronic rodent controls, pest controls, traps etc.

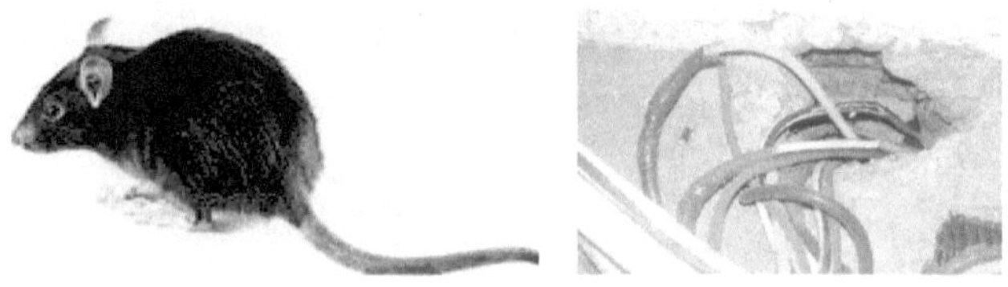

Photo by Author

Summary

- Ensure rodents are not present in premises.

- For this use electronic rodent repellant or rodent stick pads.

- Plug the holes from where wires are going.

- Remember rodents can chew wires and which will cause short circuit.

25) EV Fire

Fires in electrical vehicles (EV) causing fires happening in residential or commercial complexes is new threat. Off late fire incidences related to EV are observed in two wheelers and four wheelers. Such incidences happened near residences, in parking lots or on road. These being battery operated vehicles, fire happening into it are related to electrical fire. If we understand why fire incidences are happening then we can handle to avoid it. For this one can see carefully electrical involvement into it and understand the parameters of electrical component then these fires can be preventable.

There are three types of electrical vehicles viz. (1) Battery electric vehicles-these are battery operated vehicles. Batteries are charged by external electrical power. (2) Plugin Hybrid electric vehicles- these cars have both electric battery and combustion engine. Batteries are charged by external electrical power. (3) Hybrid vehicles- these have electric battery and combustion engine. Batteries are charged by combustion engine. As such mainly first two needs external charging power. Battery stacks need charging initially. Then during EV running, batteries will give power to the motor to run, means batteries will get discharged. And again, batteries need charging once partially discharged. As such four main components can be focused to prevent fires in EV viz. charger, battery, cable connecting vehicle and supply point and supply itself.

While understanding EV charging one has to keep in mind that there is AC charging and DC charging facility as shown below-

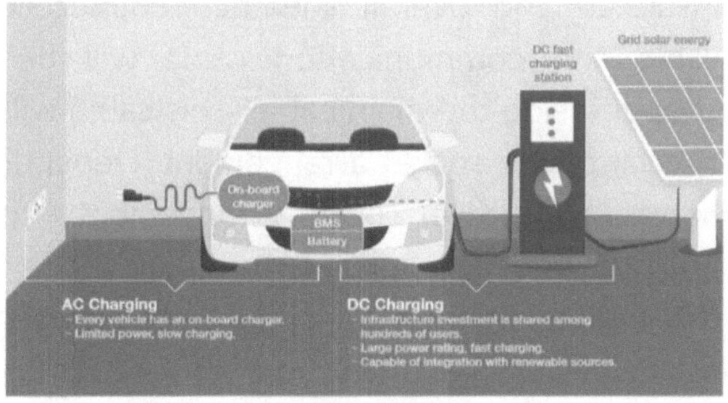

Image by Legrand

In case of AC charging, power is being taken from residential distribution board or from charging station and same is connected to AC charger available in vehicle. Where as in case of DC charging, DC power is being taken from charging station and same is connected to set of Battery through Battery management system (BMS) available in vehicle. BMS takes care of safety of charging discharging system and battery for temperature, voltage, current, thermal management of batteries, isolation between battery and chassis, optimum charging current, battery health, communication etc.

Batteries used in these vehicles are Lithium Ion and they are technically having superior performance. Lithium Ion batteries are used in EV because of their high power to weight ratio, stable at higher temperatures, fast discharge, long life and partly recyclable than other type of batteries.

Cable connected between EV and power supply should be oil and UV resistant 105 deg halogen free insulated and overall sheathed copper conductor cable and communication cable will be braided copper inside the main cable. This cable will be with special power socket arrangement. Depending upon battery kw, power socket arrangement is provided. For two wheelers 3pin 16 Amp plug top is used, for three wheelers 3pin 32 Amp plug top is used and for four wheelers generally 5pin 40 / 63 Amp and two communication port plug top as per IEC 61851 is used. In extreme cases current rating will increase depending upon manufacturer of vehicle.

Above cable tops are connected to domestic, industrial socket or charging stations connects charging gun to vehicle.

Knowing the components involved let us understand the causes of fire.

During charging of battery temperature, voltage, current, thermal management of batteries needs to be automatically controlled by system. If there is poor thermal management system then battery can lead to overcharging and increase in temperature. It will then turn into thermal runaway and thereafter explosion. Overcharging can cause fire because of internal short circuit and hence recommended charging with efficient BMS or charger controlling each cell temperature, current, voltage and health will make sure overcharging is not done. Similarly high temperature can cause fire and hence in case of high temperature of batteries the charger should not allow charging to happen. Manufacturer must ensure in process automatic switch off around 45 deg C. And as such uncontrolled charging current, faulty charger or BMS will lead in explosion.

Battery supplier should be properly selected for quality and performance of batteries before putting in to service. Else poor quality will lead to fire since it is stored energy and any defect can cause short circuit. Before installation quality check for battery leaking, damages, container, terminals etc. should be thoroughly checked. Arrangement of storage of batteries and installation procedure in vehicle should be absolutely proper, holding them firmly electrically and mechanically else during ride or in collision the mechanical parts will get loosen and will cause damage to batteries and can cause explosion. So also, battery space should properly cool with temperature sensor to give alarm in case of increase in temperature. Poor battery design, terminal arrangement can lead to short circuit as battery cells are closely placed with a separator. In case of quality or design issue of battery terminals can get shorten causing short circuit and then fire.

Present chargers used mainly are of 16 Amp and 32 Amp. Means current at rated full 16/32 A is flowing through circuit and hence power socket, distribution board, meter and connection should be adequate to carry this current. It is recommended to use RCCB at power outlet for safety. In case anyone is underrated then local heating is bound to happen and then fire. As such better to check connected load, distribution thoroughly to avoid fire incidences. Similarly new charging stations are getting installed in existing parking and hence facility engineers should check the load and to make sure power is available.

Designing selection of electronics and selection of vital parts such as batteries, charger, cables etc.

should be at climatic conditions where vehicle will be used.

Respect the battery life and change after end of life else there can be issues of overheating.

Above battery stack there are seats and design should be such that both these compartments should be fully isolated else any conducting material, liquid falling on battery will cause short circuit.

If we select good quality materials, make good design and carry out process properly then fires will not happen in EV.

Summary

- Never overcharged the batteries.
- Keep Battery compartment clean as provided by manufacturer.
- Ensure no leakage of liquids from seat.
- Use proper plug top for charger.
- Check the meter capacity and if required augment meter capacity.
- Use charging cable given by supplier.
- Keep same plug top provided by supplier and do not change the same.
- Act on the warning signals given in vehicles.

History shows that electrical fires have broken out because of one or multiple reasons cited above and due to negligence, carelessness, disregard to rules at workplace. Sparks and heating can be viewed as early indications of fire. Paying heed to these warning signals and taking prompt action to avert mishaps is duty of every person. It will save us from tremendous loss to lives and wealth.

TESTING
Comprehensive Testing

'If there is comprehensive checking in pre-commissioning then there will be less electrical faults during use.

Days has gone when insulation resistance was to be done for panels, cables, relays were to be tested, overall safety checking is done and then commissioning of MV installation was to happen. Subsequently new standards came in force, experience has taught engineers and likely faults were taken care of or best procedures were adopted making installation more and more healthy.

In spite of the developments, there were fire incidences, electrical failures and electrocution. But that do not stop engineers to plug the holes. More and more stringent testing checking are getting adopted to prevent such incidences.

Pre commissioning check is checking and testing of entire electrical system which is designed, installed as per relevant standard and is ready to be commissioned. Main focus is towards electrical fires and mainly happening in residential, commercial buildings following testing to be done and to be repeated as per frequency.

Pre-commissioning tests, checks are consolidated from various standards such as NBC, NEC, IEC 60364, CEAR Rules which specify the details. Apart from these checks now a days many additional tests are also carried out in practice.

Still it is observed that many times working of the installation in its entirety   is not verified with integral testing .and latest practices are not adopted.

It means individual components may be conforming to their respective specifications, but may fail in their stipulated performance when entire system is operating as a whole. This underscores the need to validate desired working of entire system with specified performance of the individual components working in integrity (together). Few of such examples are given below-

1) **Checking of switchgears –**

Recently It is observed of late, that although reputed manufacturers have supplied components such as MCCB, MCB, RCB after testing and verifying in factories, when installed in panels or at site, they do not operate as per desired operating curves for some or the other reason. Hence needs re-verification, testing when they reach to panel manufacturers or at site. It could be testing of tripping for overload, short circuit for tripping as per required time limits. Hence recommended to recheck switchgears and critical components at least, during assembly or after installation.

2) **Earth loop impedance-**

It is to verify that, if circuit impedance is low enough to allow sufficient current to flow when a fault occurs in an electrical installation, the fuse or circuit breaker protecting the faulty circuit should operate within a predetermined time. This was not to be tested earlier and it was only assumed that switchgear will trip if fault occurs. This test is now adopted by standard and is now widely done by many. Of course, many users do not still carry it out. By performing above test, it is ensured that any fault occurring at switch socket level will not trip back up MCB till loop impedance is in limit.

3) **Discrimination Study-**

Selection of switchgear is important from service point of view in normal condition and also in fault condition. Fault condition can be overcurrent, short circuit as well as earth fault. During fault condition there are increased electrodynamic forces and thermal effects. So, one has to ensure the switchgear selected will not get damaged and function properly as intended. While doing so upstream protective device should take care of fault and no other protective devise should operate. Hence coordination between devises is required to be done.

Selection is carried out on basis of following-

Current based selectivity- In this, setting is done for current of switchgear. Setting is such that downstream switchgear has lower setting where as upstream switchgear has higher setting than downstream switchgear. In this selectivity can be total discrimination or partial. This will depend upon particular conditions of switchgear sizes and fault currents.

Time based selectivity- In this setting is done by adjusting the time-delayed tripping units. Setting is such that downstream relay will have shortest operating time and upstream relay will have longer delays. So downstream will trip faster than upstream switchgear.

Energy based selectivity – While doing this setting, time versus current curves of upstream as well as downstream switchgears are superposed. Principally arrangement will be such that when short circuit current is detected by upstream as well as downstream switchgears then contacts of both switchgears will open simultaneously. With which current will get limited. Arc energy of downstream switchgear will enable to operate the same but not upstream switchgear.

Logic based selectivity- This setting can be done where switchgears are having electronic trip units having short time protection and ground fault protection. This is necessarily to be done along with manufacturer considering their tables.

As there are array of switchgears in a single line, there is a need to check whether a particular switchgear trips for fault occurring on that feeder and entire panel does not trip. Still in numerous installations same test is not adopted yet.

4) **Arc Flash Study**

Arc flash formation can happen when electrical current passes through air from one live conductor to another live conductor or from live conductor to ground. Due to which air gets ionized. Normally air acts as insulator under normal temperature but becomes conductive path of plasma when it becomes ionized. Due to conductive plasma path, there is explosion forming radiation, thermal energy, pressure wave and separation of hot particles. Temperature can go as high as 20000 degree K / 35000 degree F. This temperature is four times hotter than the surface of the sun.

Arc flash produces an extreme burst of bright light and molten metal disintegrates and fly in any directions. There could be many reasons for arc flash such as –

1) Loose connection

2) Dust, Corrosion, breakdown of insulation

3) Static electricity

4) Live equipment exposed to water, liquid

5) Failure of switchgear

6) Accidental touch of tools, testing probe

7) Reduction of distance between phases or phase and ground

8) Poor maintenance

Due to arc flash there could be – Burns, blindness, lung damage, ear injuries, wounds, fatal injuries for workmen and damage to equipment.

Standards available for Arc Flash are -

- IEC 62271-200 is known as the international switchgear standard, valid for 1 kV up to and including 52 kV, which specifies internal arc testing for the switchgear and control gear. It covers list of locations where faults are most likely to occur, and explains causes of failure and possible measures to decrease the probability of faults.

- IEEE 1584-2018 gives calculation methods for arc flash energy. It includes an empirically derived model for enclosed equipment and open lines for voltages from 208 V to 15 kV, and a theoretically derived model applicable for any voltage.

- NFPA 70E gives electrical safety-related work practices, safety related, maintenance requirements and other administrative controls for the practical safeguarding of employees. It also includes an informative guide on selection of arc-rated clothing and other PPE.

Study is carried out by software for single line study, short circuit current, evaluation of fault, safe working distance, flash protection boundary.

In short it is advisable to carry out arc flash study for safety. Example of Arc flash happening in panel due to loose connection-

Arc flash in panel due to lose connection at switchgear

Image by Author

5) **Thermal scanning-**

Once installation is commissioned wherever there is a loose connection, hot spot starts getting developed and as such at early stage one can take action to verify if there is loose connection or overloading or any internal fault happening. This is to be done after commissioning and also on yearly basis so that one can know the likely failures in advance. Action can then be initiated immediately to avert fire incidence.

A Thermal Camera by Fluke. Image from www.fluke.com

6) **Earthing-**

Earthing is provided but whether it is continuous or not, if equipotential is done or not are the things to be verified. Hence checking of following is must-

- Soil resistivity
- Earth pit resistance
- Total earthing value
- Earthing sizing
- Earth continuity checking
- All earthing pits to join together at ground level
- Equipotential bonding if done
- Welded / brazed joints to paint
- Earth pit marking

- Earthing points to be marked with sticker

- Earth loop impedance testing

- If earthing connection done at desired earth point of equipment

- GI/Copper as designed if used

- Earthing joints checking for use of nuts, washers

- Connection tightness

Above are only few tests. In addition, many new tests are being adopted. Same are being added in proposed testing check list given below-

Way back in 1956 when Indian Electricity Rules were prepared initially, forms were developed which are quite useful for consumers, contractors, electricians till today by and large. They can check, inspect installation and verify themselves, if installation is in line with rules and if it is safe. But after years, lot of development has taken place and new techniques / standards evolved underlining the need for fresh pre-commissioning tests to be drafted. In fact, many specifiers / users are preparing their own forms and they are being used.

On basis of this consultants / contractors should prepare a comprehensive script in which list of panels, distribution boards, cables, wires, switch sockets etc. along with their tests will be described. So that each and every electrical component will be thoroughly checked and will not be missed as unattended.

In view of all above for **medium/low voltage Installation** following form can be also considered while checking-

	PRE-COMMISSIONING CHECK LIST	REMARK
	Medium / low voltage Installation	
	PART I	
1	Report No.................	
2	Date of Inspection..............	
3	Voltage and system of supply:	
	(i) Volts............................	
	(ii) No. of phases......................	
	(iii) AC/DC...........................	
4	Name of the consumer/owner...............	
5	Address of the consumer/owner.............	
6	Location of the premises.....................	
7	Particulars of the Installation:	
7.1	Motors:	
	Make No. H.P Amps. Voltage	
	(1).........................	
	(2).........................	
	(3).........................	
	(4).........................	
	(5).........................	
7.2	Other equipment (complete details to be furnished):	
	(1)..........................	
	(2)..........................	
	(3)..........................	
	Total connected load h.p/kva.......................	
7.3	Generators (in the case of consumer himself generating energy):	
	(1)...........................	

	PRE-COMMISSIONING CHECK LIST	REMARK
	(2)...........................	
	(3)...........................	
8	General condition of the installation:	
Rule	Indian Electricity Rules, 1956 & Requirements Reports	
3	Is the list of authorized persons properly made and kept up to date duly attested?	
29	(i) Is/Are there any sign(s) of over loading?	
	(ii) State if any unauthorized temporary installation exists.	
	(iii) Are the electric supply lines and apparatus so installed, protected, worked and maintained as to prevent danger?	
	(iv) Any other general remarks.	
30	Service line and apparatus of the supplier on consumer's premises.	
	Give report on condition of service lines, cables, wires and apparatus and such other fittings placed by the supplier/owner on the premises.	
31	Has the supplier provided suitable cut-outs within consumer's premises, in an accessible position? Are they contained within an adequately enclosed fire proof receptacle?	
32	(i) State if switches are provided on live conductors.	
	(ii) State if indication of a permanent nature is provided as per this rule so as to distinguish neutral conductor from live conductor.	
	(iii) Whether a direct link is provided on the neutral in the case of single phase double pole iron-clad switches instead of fuse?	
33	(i) State if earthed terminal is provided by the supplier.	

	PRE-COMMISSIONING CHECK LIST	REMARK
	(ii) Is the Consumer's separate earth efficient?	
	(iii) Report on the efficiency of the earthing arrangement.	
34	(i) Are bare conductors in building inaccessible?	
	(ii) Whether readily accessible switches have been provided for rendering them dead?	
	(iii) Whether any other safety measures are considered necessary?	
35	State if "Danger Notice" in Hindi and the local language of the district and the type approved by the Electrical Inspector is affixed permanently in conspicuous position as per this rule.	
38	State if flexible cables used for portable or transportable equipment covered under this rule, are heavily insulated and adequately protected from mechanical injury.	
44	(i) State if instructions in [English or Hindi and the local language of the district and where Hindi is the local language, in English and Hindi] for the restoration of persons suffering from electric shock have been affixed in a "conspicuous place."	
	(ii) Are the authorized persons able to apply instructions for resuscitation of persons suffering from electric shock?	
49	Leakage on premises:	
	State insulation resistance between conductors and earth in Megaohms.	
50	(i) Whether a suitable linked switch/ circuit breaker is placed near the point of commencement of supply so as to be readily accessible and capable of being easily operated to completely isolate the supply?	

	PRE-COMMISSIONING CHECK LIST	REMARK
	(ii) Whether every distinct circuit is protected against excess energy by means of a suitable circuit breaker or cut-out?	
	(iii) State if a suitable linked switch or circuit breaker is provided near each motor or apparatus for controlling supply to the motor or apparatus.	
	(iv) State if adequate precautions are taken to ensure that no live parts are so exposed as to cause danger.	
51	(i) State the condition of metallic coverings provided for various conductors.	
	(ii) (a) State whether clear space of 90 cm is provided in front of the main switch board.	
	(b) State whether the space behind the switch board exceeds 75 cm in width or is less than 23 cm.	
	(c) In case the clear space behind the switchboard exceeds 75 cm. state whether a passage way from either end of the switchboard to a height of 1.80 meters is provided.	
61	(i) Has the neutral conductor at the transformer been earthed by two separate and distinct connections with earth.	
	(ii) Have the frame of every generator, stationary motor and so far as practicable portable motor and the metallic parts (not intended as conductors) of all transformers and any other apparatus used for regulating or controlling energy and all medium voltage energy consuming apparatus been earthed by two separate and distinct connections with earth?	

	PRE-COMMISSIONING CHECK LIST	REMARK
	(iii) Have the metal casings or metallic coverings containing or protecting any electric supply line or apparatus been properly earthed and so joined and connected across all junction boxes as to make good mechanical and electrical connection'	
	(iv) State if the consumer's earth electrode is properly executed and has been tested with satisfactory results.	
	(v) Is the earth wire free from any mechanical damage?	
	Overhead Lines:	
74	(i) State if the consumer has any overhead lines and if so their condition to with specific reference to relevant rules.	
93	(ii) Is there any other overhead line near the premises of the consumer which does not comply with rule 79 or 80?	
	(iii) Is guarding provided for overhead lines if it is inside the factory, for road crossings and busy localities?	
	(iv) Any other remarks.	
	Part II	
1.	As built drawings to be made available at site approved by designer.	
2.	Environmental data such as date, time, temperature, humidity of that day and earlier day to note.	
3.	List of test instruments to make and to check make of instrument, calibration certificate, calibration authority, certificate validity date.	

	PRE-COMMISSIONING CHECK LIST	REMARK
4.	Visual Inspection checks for installation including-	
5.	Visual inspection will include if installation as per design, safety against electric shocks, checking as per standard and checking clearances measurements. During inspection if any alteration is required to be done then to check it will not compromise for standard. To check following as per drawings- a) Cable sizing b) Panels switchgear c) Distribution boards switchgear d) Switch sockets e) Lighting f) Wiring Add components as applicable as per drawing.	
6.	To check installation as per drawing from Meter room, Shafts, Electrical Room, Electrical closets, Distribution Boards, Switch sockets, cables, wiring and earthing if as per drawing.	
7.	Electrical shafts will not have other services, pipes etc. to confirm	
8.	Battery room will be separate.	
9.	Electrical installation will be clean, dust free and away from water.	
10.	To check fire barriers and precautions against fire and protection against thermal effect are being taken in shafts, electrical rooms, panels etc.	

	PRE-COMMISSIONING CHECK LIST	REMARK
11.	Checking of conductors as per current carrying capacity and load. This will also include for voltage drop.	
12.	Verification of isolating and switching devices correctly located and selection as per external parameters.	
13.	Neutral and protective conductor identification.	
14.	Checking and settings of protective and monitoring devices such as switchgears, relays for overcurrent, short circuit. Panel 1- Panel 2- Panel 3- Etc.	
15.	Accessibility for convenience of operation and maintenance.	
16.	Presence of drawings, diagrams, warning notices.	
17.	Discrimination study of switchgears in panels. Panel 1- Panel 2- Panel 3- Etc.	
18.	Checking of switchgear ACB, MCCB, MCB, RCD for tripping by third party agency to ensure if they are tripping properly as per curve or not and to doubly check if there is manufacturing defect. Panel 1- Panel 2- Panel 3- Etc.	

	PRE-COMMISSIONING CHECK LIST	REMARK
19.	Usage of plain washer, spring washer, tightening, torque application and marking on cable, busbars terminations. Usage of bimetallic washers wherever required. Panel 1- Panel 2- Panel 3- Etc.	
20.	If single compression / double compression cable glands are used and they are earthed. Panel 1- Panel 2- Panel 3- Etc.	
21.	All holes in panels, junction boxes are closed.	
22.	No loose jointing is done and lugs are used everywhere.	
23.	All blocking materials that are used for safe transportation in switchgears, contactors, relays, etc, are removed	
24.	All connections to the earthing system are feasible for periodical inspection like providing test links.	
25.	To check if sharp cable bends are absent, using suitable support clamps at regular intervals with marking and arrows showing flow of power.	
26.	Circuit breaker or lockable push button is provided near the motors / apparatus for controlling supply.	
27.	Two separate and distinct earth connections are provided for the motor/apparatus / panels, DBs, conduits etc.	

	PRE-COMMISSIONING CHECK LIST	**REMARK**
28.	The metal rails on which the crane travels are electrically continuous and earthed and bonding of rails and earthing at both ends are done.	
29.	All equipment are weather, dust and vermin proof	
30.	Emergency lighting is placed in important rooms such as electrical ups and or any other closed rooms, evacuation passage. Load testing on batteries and UPS. And back up time of 90 min is available. Room 1- Room 2- Room 3- Etc.	
31.	Exhaust fans, ventilation is provided in above rooms and in working condition.	
32.	Wherever AC is needed same is provided and in working condition especially UPS, Battery Rooms.	
	Part III - TESTING	
33.	Separation of live parts from other circuits and from earth shall be verified by measurement of insulation resistance.	
34.	Electrical continuity and conductivity tests of protective, equipotential and earth bonding conductors, earthing pit results, combined earth grid results.	
35.	Resistance of whole lightning system to earth	
36.	Continuity of protective conductors, equipotential bonding by supply having no load voltage of 4V to 24 V dc / ac with minimum current of 0.2 A.	

	PRE-COMMISSIONING CHECK LIST	REMARK
37.	Insulation resistance tests between live conductors and the protective conductors connected to the earthing arrangement for cables, busbars etc.	
38.	Insulation resistance of electrical installation between live conductors Phases, live conductor Phases and Neutral, live conductor Phases and Earth, Neutral and earth. Minimum value of insulation resistance will be greater than 0.5 MΩ (1 MΩ as per IE Rule) for dc test voltage of 500 V for nominal circuit voltage of up to and including 500 V.	
39.	In case of non conductive floor and wall, Floor and wall resistance is measured between protective conductor and test electrode in floor or wall comprising metallic plate with sides 250 mm on which 270 mm damp water paper / cloth is placed and force of 750 N or 250 N is applied on plate as per standard.	
40.	Verification of condition for protection by automatic disconnection of supply.	
41.	Measurement of earth loop impedance, verification of characteristic of associated protective devices such as MCCB, MCB, RCD for overcurrent, short circuit, earth leakage.	
42.	Polarity test to be carried out for especially single-phase equipment connected in phase, at change over switch and at socket if phase neutral earth are connected properly and as required.	
43.	Functional test will be carried out in panels for switchgears, load, controls, interlocks, protective devices, relays etc. to check installation as required.	
44.	Voltage drops checking for cables.	

	PRE-COMMISSIONING CHECK LIST	REMARK
45.	In case any test is failed then to repeat after fault rectification	
46.	Test of compliance of SELV (Safety Extra Low Voltage) and PELV (Protection by Extra Low Voltage) circuits or for electrical separation	
47.	Check of phase sequence in case of multiphase circuit	
48	Functional test of switchgear and control gear by verifying their installation and adjustment	
49.	Thermal scanning after initial charging and checking if any hot spot are present and if so to attend loose connection or overload if happening.	
50.	Conformity assessment (with standards and specifications) of equipment used in the installation	
	PART IV EQUIPMENT ASSESSMENT	
51.	The conformity assessment of equipment with the relevant standards can be attested: By mark of conformity granted by the certification body concerned, or By a certificate of conformity issued by a certification body, or By a declaration of conformity given by the manufacturer	
52.	Marking and Conformity It means that product can have following check list so that product guarantee will increase: The product meets the standard requirements such as IEC / IS.	

	PRE-COMMISSIONING CHECK LIST	REMARK
	Conformity to use product for particular usage such as Marine, Nuclear application etc.	
	Quality assurance by way of Type testing certification availability.	
	Routine test certification	
	Quality assurance certification like ISO certification.	
	Third party verification of samples.	
	Part V DOCUMENTATION	
53.	Documentation verification- Type Tests Certification Factory Test	
	Site acceptance test Drawing verification Handing over document verification	
	Part IV CLEARANCE	
54.	To give clearance for commissioning	

While enumerating the list, High Voltage and Extra High Voltage is excluded and considered only for Medium voltage, since there are standards available for testing of equipment involved in EHV and HV such as Transformers, HV Panels, HV cables, Motors, Generators etc.

Periodic check-testing of an installation

In many countries, all industrial and commercial-building installations, together with installations in buildings used for public gatherings, must be re-tested periodically by authorized agents.

The following tests should be performed not limited to are-

- Verification of Relays, RCD effectiveness and adjustments

- Appropriate measurements for providing safety of persons against effects of electric shock and protection against damage to property against fire and heat

- Confirmation that the installation is not damaged

- Identification of installation defects

The table below shows the frequency of testing commonly prescribed according to the kind of installation concerned.

Type of installation		Testing Frequency
Installations which require the protection of employees	Locations at which a risk of degradation, fire or explosion exists	Annually
	Temporary installations at worksites	
	Locations at which MV installations exist	
	Restrictive conducting locations where mobile equipment is used	
	Other cases	Every 3 Years
Installations in buildings used for public gatherings, where protection against the risks of fire and panic are required	According to the type of establishment and its capacity for receiving the public	From one to three years
Residential	According to local regulation	From one to five years

As such pre-commissioning checking if done thoroughly and inspection is carried out properly then the installation will be healthy and no fire risk will be involved.

In case modification is done in installation it is recommended to re check the installation.

Again, checking is to be repeated annually or as stipulated and proper precautions are taken then one can say installation will remain healthy for years.

SAFETY
Safety Related To Electrical Fire

Safety rules are numerous and already in practice as per organization to organization and likewise mentioned in all standards worldwide and also rules are formed. Apart from standards, local government acts, guidelines, rules, no objection certificates issued also are to be verified thoroughly while design, planning and implementation. In all these documents guidelines for safe working, fire prevention, life safety is provided.

As our purpose is to exterminate electrical fires then some safety rules play major roles by default. By just adhering to safety, fires can be avoided. Such pointers, highlights are given below which are good enough to prevent many electrical fires. But they are either not accessible or not observed at local electrician level or in small projects. In such cases the rules given below will be beneficial.

Safety related with electricity is comprehensive, covering many aspects and having multi faced dimensions from planning, design, execution, testing, maintenance till usage. Salient points in procedures and prevention are collated and given underneath and should be remembered.

The rules given below are very easy, generic and self-explanatory. Users must strictly follow rules & regulations stipulated in standard. At the same time to avert electrocution, unusual blaze, accident one must have quick overview of installation and hence following safety tips will be useful in practice.

A) **PROCEDURES**

1) While working on live mains, permit-to-work system must be followed. In each organization permit system is present in which work to be done, authorized to do work, supervisor etc. details are present. This system is to be extensively used so that safety officer is aware where works are going out and accordingly deployment of safety supervisor can be done at that site. This may not be true in case of small house work but owner or the electrician must know or get to know what is intended work and accordingly which precautions are to be taken. He must follow them and remain attentive.

2) Instructions related to operation of switches shall be recorded in the register. When there is major modification or work is to be carried out for which it is necessary to put off power and then further work is to be done then LOTO lock out / tagout system is to be used for safe working. This process is carried out in large installations and having multiple panels and equipment interlinked to each other. But this is the safest process from work safety perspective. At

the same time many a times over confidence is seen by users or workers while working to keep mains off which is improper and one should put off the supply and then only work.

3) Only competent, experienced and authorized persons shall work on live mains & apparatus. Electrical work is to be carried out by electrician, supervisor and contractor who have licenses. While doing any small or large scale work it is responsibility of customer to verify if they have valid licenses.

4) While working, sound gauntlets, rubber mats, tools, platforms or other accessories and safety devices are to be used. Any short cuts will be harmful for person who is working as well as for installation. While welding wear goggles, safety glasses or any other eye protection as instructed by the person in charge depending upon the type of work handled.

5) While observing near live part minimum working distance from the exposed live high voltage mains and apparatus will be maintained. Sometimes observation, readings are required to be taken in substation. At that time utmost safety is to be taken as prescribed by organization.

6) High voltage contact indicator & phasing rods are to be used while working on high voltage and discharged to earth after making them dead duly earthed before doing any work. This is applicable in substation or on transmission lines, when outages are

taken for work. In such cases supply is cut off and locking to switchgear is done but before proceeding, verification of stray supply is checked up and also phase shorting rod is used so that any accidental on will trip instantaneously.

7) All earthing switches wherever installed should be locked up. While working on high voltage panels once breaker is made off and supply is cut off, wherever applicable the switch trolley is to be withdrawn and earthing trolly is inserted and made on. So that phases are completely earthed. With this any accidental on, will trip total system because of earthing. Hence earthing switch should be applied and further it should be locked so that it remains secure.

8) Calibrated tools and tools with proper insulation should be used. After usage insulation of tools get worn out or gets removed. This is applicable for tester, screw driver, pliers, test lamp, meggar which leads to high voltage testing probe. It is a big question that when insulation of such tools was checked last time It should be done regularly because in these hand-held tools supply is always passed through it.

9) While working on critical electrical equipment fire extinguishers should be kept near and accessible where work is going on especially where there is hot work going on such as welding. Already we have discussed this point at length.

10) 'Danger' sign boards are displayed at conspicuous place on the mains and also on the equipments and locations such as substation, transformers, panels, distribution boards in especially local language, Hindi, English.

11) Working place should be well illuminated and ventilated so that people who are working will see the object, exit path properly and so also any gases formed will pass out of the place and will not suffocate the workers.

12) While working especially on electricity, clothes preferably should not have metal buttons, metal straps and similar metal fittings as they can conduct electricity. So also, loose clothing should not be worn which might get stuck up in any object. Firstly, working on live conductor or on panels are prohibited but sometimes electrician has to work where apparatus is off but supply is present. On such occasion rule is not to roll up sleeves as dry cloth gives some protection against shocks and also not to wear arm suspenders, arm bands with metal buckles or other metal parts. At the same time metal key chains or metal keepers for key rings or watch chains should not be worn on outer side of clothing. Treat all electrical conductors and apparatus always as live. All voltages are dangerous. It shall be borne in mind that even low voltage shock may be fatal.

13) Electricity and water are conflicting elements and should not be brought near each other directly or indirectly at any time. Therefore, working area, tools, hands, equipment should be dry while work is going on. In case one has to work in damp area like pump room then extra precautions should be taken like putting off supply and drying all the equipments and then to work.

14) Ensure two supplies are not mixing causing short circuit and large spark. This is applicable when mains and DG supplies are coming in change over switch or while applying test voltage on any live mains. As such mains are to be put off, get permit to work and then start further work.

15) Situation arises when old installation is present but mains is made dead and now time is there to connect the same to live mains. In such case live mains are to be connected last after checking existing unused installation. Also, it is required to ensure no mixing of phases, phase sequence and that installation is healthy.

16) Work might be small or big but earthing and switchgear protection will safeguard always and hence keep in mind not to ignore this before doing work.

17) While working on any cable such as High voltage or Low voltage for cutting or any operation such as removal, rerouting shall be identified first and ensure that same is dead.

18) After testing the high voltage cable with meggar, the cable shall be discharged through a 2 meg ohms resistance and care will be taken not to work directly.

19) On completion of work, removal of the earthing and short- circuiting devices shall be carried out in the reverse order to that adopted for placing them.

20) No one should face arc or flash but cultivate the habit of turning your face away from arc or flash whenever it may occur with mouth closed.

21) Before working in manholes, closed underground trenches, shafts care is to be taken that there could be accumulation of gases. And also, prevention for not accumulation of the same to be done.

22) In meter room behind meters at some places wooden board which is varnished is used which is very dangerous since it emits flammable vapor. So also only authorized person can test, change meter, boards and cutouts. Opening of switch or inserting fuse to be done quickly and positively.

B) PREVENTION

1) In building essentially vertical opening between the floors of the building shall be suitably enclosed or protected by fire sealant.

2) HVAC System shall be provided with dampers designed to close automatically in case of fire.

3) For large places of assembly, prevention of circulation of smoke to the system may be provided.

4) Separate air handling units for various floors shall be provided.

5) Smoke venting facility to be provided in case of fire.

6) Sprinklers, smoke detectors, gas-based suppression system to be provided as applicable.

7) Surface interior finishes should be selected so that it shall not generate toxic smoke/fumes. Glazing, skylights, louvers   with adequate fire resistance and frames   to be used.

8) General exit requirement shall be maintained such as it will remain unobstructed, sufficient numbers and width, visibility, exit signage, emergency lighting and fire doors.

9) Lift and escalators   should not be considered as exits and not to be used in case of fire.

10) Fire switch to be provided for fire lift at ground level.

11) Wherever required lift enclosure will be fire rated, lift motor room will be on top of shaft having ventilation and smoke extraction vent.

12) Pressurized lift shaft to be provided.

13) Lift should have emergency switch, lamp and communication facility.

14) Fire lift will be specifically made as specified.

15) Basement level will be ventilated, mechanical smoke extractor, openable window at external wall.

16) Service duct/shaft will have fire rated doors and vent opening at top.

17) Internal staircases shall be constructed of noncombustible material.

18) Around lift shaft there will not be gas piping, electrical panel.

19) Staircases will have adequate width of tread, proper riser height, hand rails, proper head room and pressurization unless openings are provided to exterior.

20) All building depending upon the occupancy use and height shall be protected by fire pump, fire extinguisher wet riser, down comer, automatic sprinkler installation, water spray, foam, gas or dry powder system as per provision.

21) For purpose of firefighting water storage tank of adequate capacity to be provided.

22) Refuge area will be provided as required with amenities mentioned there in.

23) Lightning Arrestor to be installed on building.

24) Earthed or Earth neutral conductor will not have cutout, link.

25) Confirm that bare conductor will remain inaccessible.

26) Flexible cables shall not be used for portable machines unless heavily insulated & protected from mechanical injury.

27) Consumer must use circuit breaker for low, medium or HV consumer at entry point.

28) Circuit breaker or emergency tripping device to be placed adjacent to the motor or group of motor.

29) Switchboard shall have clear space in front as well as behind as specified.

30) Earth Leakage Protective Device to be used.

31) For Outdoor Substation Clearance to be maintained as specified.

32) Substation with oil filled transformer will have baffle wall, oil soak pit & fire rated doors & Fire extinguisher system as specified.

33) In case of high or extra high voltage suitable interlock between isolator & controlling circuit breaker, isolator & corresponding earthing switch to be provided.

34) Cable trenches will have non-flammable covers or filled with sand.

35) Installation should be free from temporary wires, consumables.

36) All equipment essentially should be weather, dust and vermin proof. And as per area selection further classification such as flame proof, water tight etc. to be made.

37) Ensure distance between phases, phases and neutral, phases / neutral and earth.

38) Nameplates, drawings are to be placed on / near equipment.

39) Installation should be checked by licensed supervisor; certificate of concerned contractor should be available and checked by electrical inspector.

DESIGN TOUCH

Electrical design is vast ocean but 'electrical fire' topic cannot be completed unless we go through vital equipment selection, design parameters which are mainly related to 'electrical fire'. This is more or less related to part of design topic. Here theory part of design is excluded and concentration is only on selection parameters of important equipments are mentioned. This will help in order to be familiar with the advantages they are associated with. While selecting the equipment, focus will be on our subject of 'electrical fire'.

After going through statistics of electrical fire, it is observed that-

- Maximum electrical fires are happening in Residential buildings.

- Electrical fires are happening due to short circuit.

- Apart from above there are human mistakes such as not doing maintenance, improper testing, un awareness, not adhering to rules etc.

- Also, natural disasters like lightning or hurricane.

Following table will indicate where and why the fires took place in past.

Sr. No.	Items	Residential buildings	Short circuit	Human mistakes	Natural disasters
1	Meter room	Y	Y	Y	Y
2	Electrical shaft	Y		Y	
3	Earthing	Y	Y	Y	
4	Wiring	Y	Y	Y	
5	Cables, Wires	Y	Y	Y	
6	Switchgears	Y	Y		
7	Lightning Arrestor				Y
8	Distribution boards	Y	Y	Y	
9	Switch sockets	Y	Y	Y	

Taking into consideration above causes and places, now let us discuss one by one. It will be appropriate to focus on most important equipment and topics which are given below-

- Meter room
- Electrical shaft
- Switchgears
- Cables, Wires
- Distribution boards
- Wiring
- Switch sockets
- Earthing
- Lightning Arrestor

METER ROOM

1) Selection-

   1) There should be a separate meter room in the premises exclusive for incoming cables, meters etc. And should not be closet created in passage or cupboard on outer wall.

   2) Sufficient space to accommodate meters

   3) It should be well ventilated

   4) Unobstructed working distance of minimum one meter from meters.

   5) Should not be near to equipment generating vibrations and magnetic fields.

   6) Level of meter room should be above flood level

   7) Absence of water seepage and meter room should not be below toilet.

   8) Must not be under staircase, preferably should be under electrical shaft

   9) It should not be in any elevated area with top cover and door.

   10) Should be at ground floor with direct access for entry to person and electrical cable.

   11) It should be free from dust, fumes, moisture, direct sunlight.

12) Meter room areas must not be prone to fire and toxic hazards.

13) The site temperature should be within the limits of 0°C to +50°C.

14) Main switch should be easily accessible to put off the supply in case of emergency.

15) Copper conductor PVC insulated wires to be used with lugs for terminations for phases, neutral and earth connection to the consumers.

16) Separate neutral to each consumer up to the metering point and same shall be used by the consumer. Neutral should not be mixed with consumers.

17) Wherever there are multiple meters' installations, busbar arrangement with copper busbars mounted on insulators in dust vermin proof panel duly earthed to be used.

18) Neutral should not be earthed after the metering point.

19) The energy service provider provides the earthing connection in meter room but it is preferred to have continuous connection with the earthing pit of the transformer neutral point.

20) Cut outs were provided with meters and are now changed to MCB. But still this practice is adopted which should be stopped.

21) Door of meter room should be preferably of fire rated openable outward.

22) Emergency light, fire extinguisher to be provided in room.

23) Internal wiring should be neat and spacious on asbestos board with cable trays above.

24) Checking of meter capacity, wiring, main cables and switch to be verified with reference to load of consumer. And in case load is increased then same to be changed as per capacity immediately.

ELECTRIC SHAFT

1) Location of shaft should be such that it will start from meter room and will be straight vertical.

2) Shaft should not be used for storage of materials.

3) Shaft should be closed on each floor by fire sealant so that smoke will not travel from meter room in to shaft or from one floor to another floor. And floor shaft to floor or common lobby.

4) Cables, conduits or bus duct should be installed on proper support and not on wall directly.

5) Cables should be neatly dressed. Cables and conduits to be neatly clamped and adequate distance between two to be maintained.

6) In case of multiple shafts, cable trays to be used in meter room for horizontal travelling of cables.

7) Electrical shaft to be used for electrical services only and for no other services.

8) Earthing to be taken from meter room throughout shaft from which earthing for users can be tapped.

9) Electrical shaft will have two hours fire rated door.

EARTHING

When it comes to electrical fire then earthing is a must to be checked if it is proper. Because improper earthing could be a reason of electrical fire. When the internal fault happens and it starts developing to earth fault then same is required to be cleared by switchgear. For which installation earthing needs to be proper as per standard.

Provision of adequate earthing is extremely important for the safety of operating personnel as well as for proper system operation. Earthing means connecting the electrical equipment to the general mass of the earth which has a very low resistance. Earthing ensures Safety of personal and equipment. It also acts as the reference to all control signals.

Earthing conductor between enclosure, interconnecting earth bar and earth mesh provides equipotential points creating equipotential platform. It eliminates all the hazardous touch potential differences.

Earthing types are classified as TN, TT and IT system.

In TN system- Neutral point which is getting generated from star point of supply source is earthed. There is protective conductor in between source earth point and the exposed conductive parts of installation. Installation is earthed by this protective conductor. So, there is a metallic path for earth fault currents to flow from the installation to the earthed point of the source through earthing in case of fault. TN systems are further sub-divided into TN-C, TN-S and TN-C-S systems. In case of TN C – N neutral and PE (Protective earth) conductors are combined (PEN). In case of TN S – N neutral and PE (Protective earth) conductors are separate. And in case of TN C S – TNS is used downstream of TNC.

In TT system — Neutral point which is getting generated from star point is earthed. The exposed conductive parts of installation are connected to local earth electrodes. It means both earthings are electrically independent.

In IT system- The source point is either unearthed or earthed through a high impedance and the exposed conductive parts of the installation are connected to electrically independent earth electrodes.

By and large usage is of TNS and TT type and hence we can overview the system from fault and type of switchgear to be used point of view. This is important because in case of fault, circuit should be disconnected. Our aim is in case of fault switchgear should trip. Following drawings will highlight on both type of systems and what happens in case of fault. IT Type is used in hospitals.

When there is dead short circuit between three phases or two phases or phases and earth then fault current is very high. This will depend upon source impedance of transformer and conductor impedance. Near the transformer secondary it will be highest as per short circuit current of transformer and will start reducing as conductor length increases. However, if there is phase to earth fault or earth fault generated because of internal short circuit then the situation changes as per earthing type which is discussed below.

Case study is considered for calculation purpose which will be applied in case of both the systems.

Transformer of 1000 KVA with 5.5% impedance, secondary 400V is provided. From transformer 3 Nos 50 Mtrs of $3^1/_2$cx400 sq. mm A2XFY cables are taken up to premises of apartment where main panel is provided. From main panel 2 Nos. 6 sq. mm + 1 No. 4 sq. mm copper conductor PVC insulated wires 20 Mtrs are provided in flat in Distribution board. In such case Impedance of Transformer will be 0.009 ohm, Cables will be 0.0022 ohm and wires will be 0.079 ohm.

Earthing pit resistance is considered 5 ohms.

In such case let us check what happens in case of fault-

TN-S SYSTEM

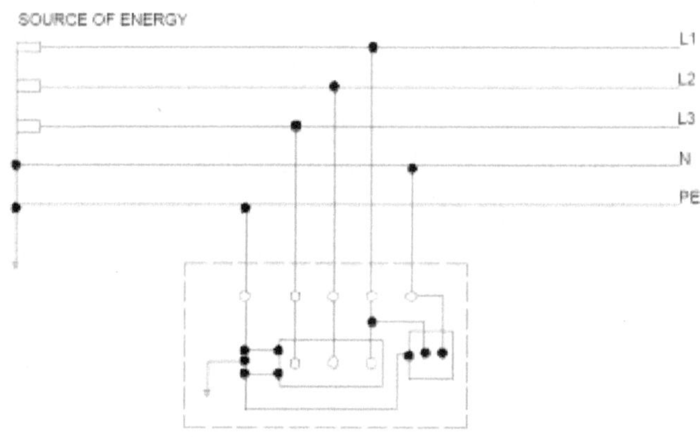

TNS SYSTEM

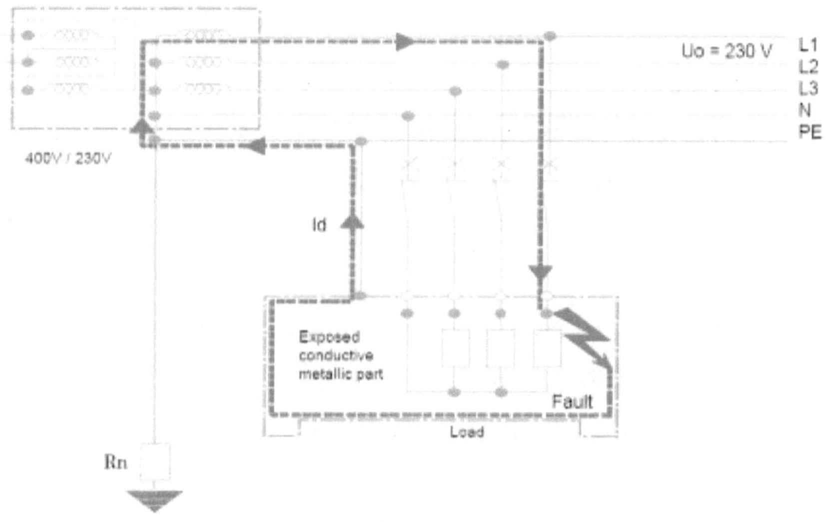

In case insulation of wire is damaged and fault has occurred. Then fault current starts from source to phase, goes to body which is earthed and back to source. In this case fault current Id is 2.5 KA calculated from (230 V / (0.009+0.0022+0.079) ohm). This is very strong current. It means if short circuit protective device (SCPD) used such as MCCB, MCB then in case of fault it will trip provided KA rating selection should be proper as per fault level. MCCB has short circuit capacity 10 times rated current. For 63 A MCCB it will trip at 630 Amp. Hence in case of fault current of 2500 Amp it will trip. And hence essentially in case of TNS system MCCB is to be used as per fault current of system. And RCD, MCB to be used for outgoing circuits.

TNS System

- Requires the installation of earth electrodes at regular intervals throughout the installation till source but protective conductor is common.

- Requires that the initial check on effective tripping for the first insulation fault be carried out by calculations during the design stage, followed by mandatory measurements to confirm tripping during commissioning.

- Modifications or extension should be designed by qualified person.

- In the case of insulation faults, greater damage may cause to the windings of rotating machines.

- Risk of fire, may be there due to the higher fault currents.

- Hence Short circuit protective devices and Residual Current devises to be used.

TT SYSTEM

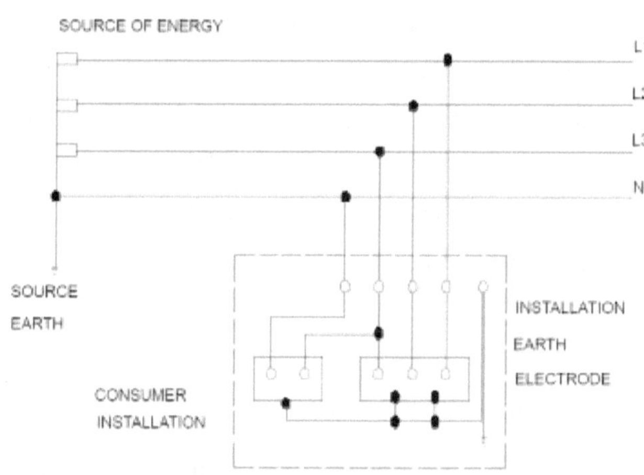

TT SYSTEM

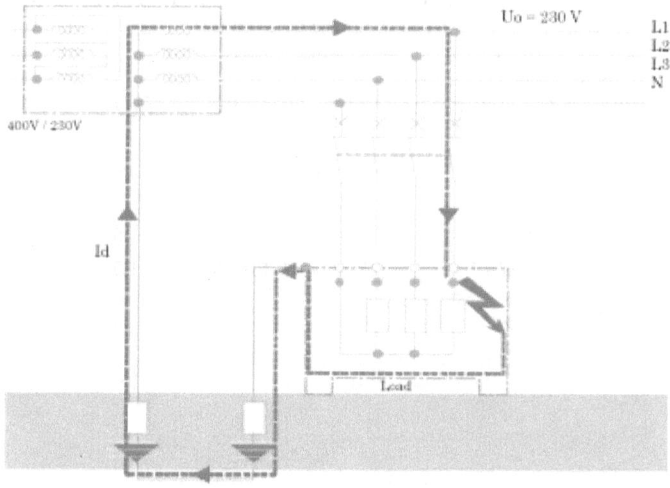

In case insulation of wire is damaged and fault has occurred. Then fault current starts from source to phase, goes to body and then to local earth, mother earth and source earth and then back to source. In this case fault current Id is 23A calculated from (230 V / (0.009+0.0022+0.079+5+5) ohm). This is very weak current. It means if only short circuit protective device (SCPD) is used such as MCCB, MCB then in case of fault it will not trip. MCB has short circuit capacity from 4 times to 10 times as per curve. For C curve 10 A MCB it will trip at 60 Amp. Hence in case of fault current of 23 Amp it will ignore. And hence essentially in case of TT system RCD is to be used along with MCCB, MCB for outgoing circuits.

TT System

- Simplest solution to design and install.

- Used in installations supplied directly by the Public LV distribution Network

- Does not required continuous monitoring during operation

- Periodic check on the RCDs necessary.

- Protection is ensured by the residual current devices (RCD).

- RCD also prevent the risk of fire when they are set to 500mA.

- Each insulation fault results in an interruption in supply of power limited to faulty circuit by installing the RCDs.

- Loads or parts of the installation which, during normal operation cause high leakage currents, require special measures to avoid nuisance tripping, i.e. supply the loads with a separation transformer or use specific RCDs.

- If exposed conductive parts are earthed at a number of points, an RCD must be installed for each set of circuits connected to a given earth electrode. Hence Short circuit protective devices and Residual Current devises to be used.

IT SYSTEM

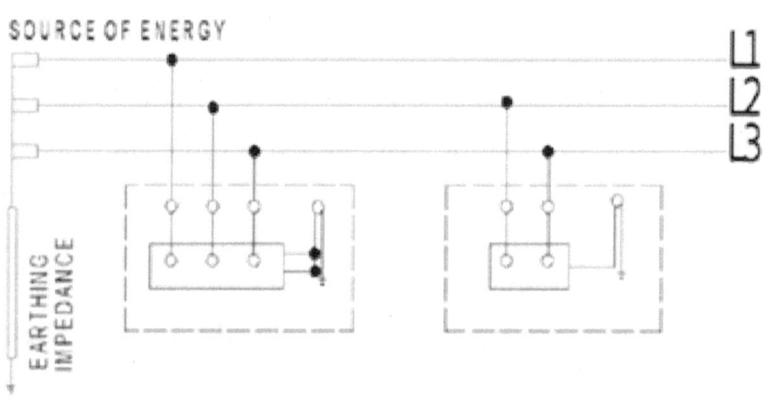

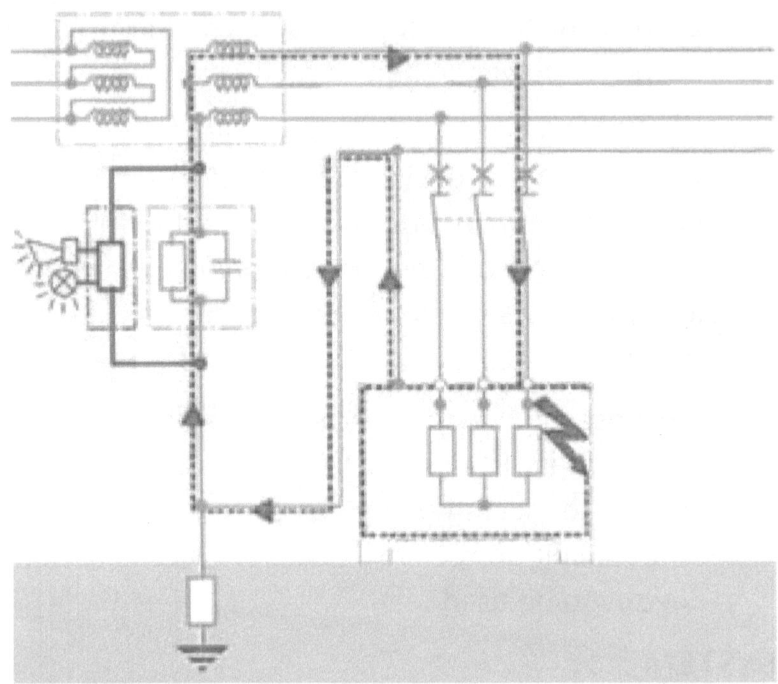

In IT system and in normal operation, the system is earthed by the earth impedance at source and exposed conductive parts of installation are connected to local earth electrodes.

In case insulation of wire is damaged and fault has occurred. Then fault current starts from source to phase, goes to body and then to local earth, mother earth and source earth and earth impedance (say 1500 Ohm) and then back to source. In this case fault current Id is 0.15A calculated from (230V/ (0.009+0.0022+0.079+5+5+1500) ohm). This is very negligible current. It means if only short circuit protective device (SCPD) is used such as MCCB, MCB then in case of fault it will not trip. Hence in case of fault current of 0.15 Amp it will ignore. And hence essentially in case of IT system RCD is to be used along with MCCB, MCB for outgoing circuits.

IT System

- Solution offering is the best where service continuity is required without interruption.

- Generally used in installations supplied by a private MV/LV or LV/LV transformer

- Requires maintenance personnel for monitoring and operation

- Requires a high level of insulation in the network.

- Protection of the neutral conductor must be insured.

- A permanent monitoring of the insulation to earth must be provided, coupled with an alarm signal (audio and/or flashing lights, etc.) operating in the event of a first earth fault.

- There is continuity of service without interruption is possible hence best to use at critical locations such as operation theaters. If there is leakage still power will not get tripped and work can be done. Subsequently fault can be attended. In such situation localized transformer LV/LV is used with IT type earthing.

TYPE OF EARTHING

1) Plate type earthing

   Copper/GI plate electrodes shall be made of 6 mm thick copper plate of 600X 600 mm size or as specified. The plate shall be buried vertically in ground at a depth of not less than 3.0 meters to the top of the plate. The pit should be filled with charcoal/

and salt in such a way that the electrode is encased to a minimum thickness of 300 mm all round. The electrode, to the extent possible, should be buried in a depth where subsoil water is present. Earth leads to the electrode shall be laid in a heavy-duty GI pipe and connected to the plate electrode with brass bolts, nuts and washers.

A GI pipe of not less than 40 mm dia shall be clamped with bolts vertically to the plate and terminated in a wire meshed funnel. The funnel shall be enclosed in a masonry chamber of 450 mm x 450 mm dimensions. The chamber shall be provided with GI frame and CI inspection cover. The earth station shall also be provided with a suitable permanent identification label tag.

Typical Arrangements of Earth Pits with Plate Type Earthing as below-

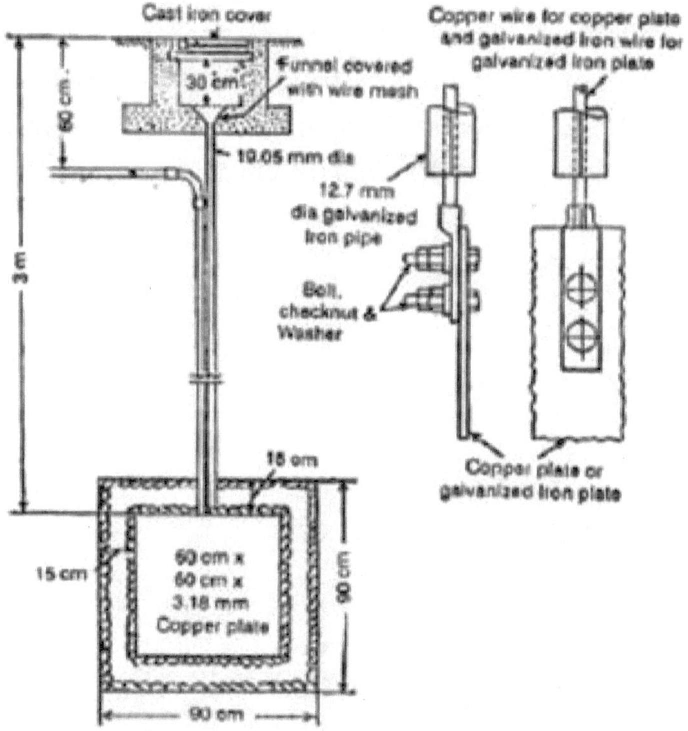

Image by Bureau of Indian Standards

2) Pipe type earthing

GI pipe of 2500 mm / 3000 mm of 40 mm diameter or Cast-Iron pipe of 100 mm diameter shall be used as electrode. The pipe shall be buried vertically in ground at a depth of 3.0 meters. The pit should be filled with charcoal/ and salt all round. Alternatively driven rods generally consist of round copper, steel-cored copper or galvanized steel 13, 16 or 19 mm in diameter from 1220 to 2 440 mm in length can be also used.

A GI pipe shall be clamped and terminated in a wire meshed funnel. The funnel shall be enclosed in a masonry chamber of 450 mm x 450 mm dimensions. The chamber shall be provided with GI frame and CI inspection cover. The earth station shall also be provided with a suitable permanent identification label tag.

Typical Arrangements of Earth Pits with Pipe Type Earthing as below-

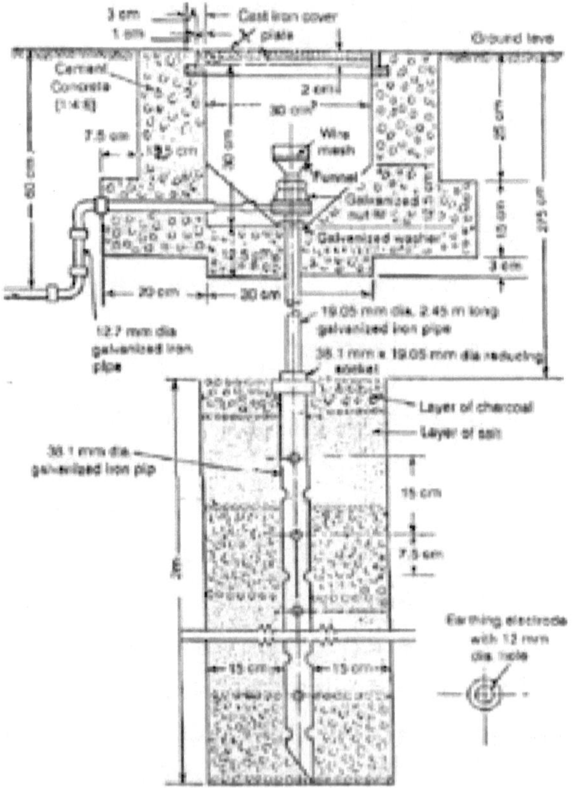

Image by Bureau of Indian Standards

Equipotential Bonding

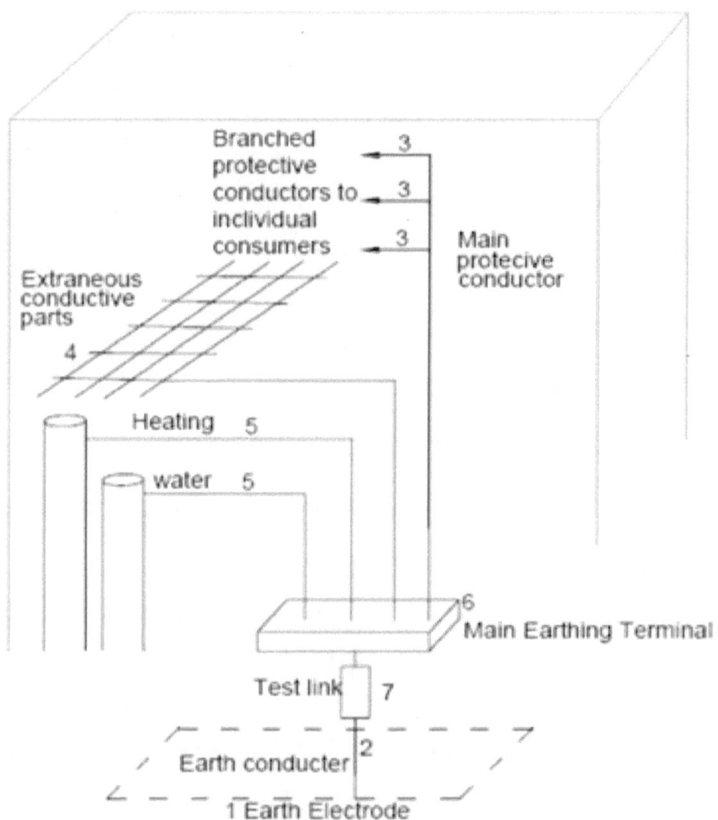

EARTH ELECTRODE (1)

A conductor or group of conductors in intimate contact with and providing an electrical connection to the earth.

EARTH CONDUCTOR (2)

A protective conductor connecting the main earthing terminal (or the equipotential bonding conductor of an a installation when there is no earth bus) to an earth electrode or to other means of earthing.

EXPOSED CONDUCTIVE PART

A conductive part of equipment which can be touched and which is not a live part but which may become live under fault.

PROTECTIVE CONDUCTOR (3)

A conductor used for some measures of protection against electric shock and intended for connecting any of the following parts:

- Exposed conductive parts,

- Extraneous conductive parts,

- Main earthing terminal,

- Earthed point of the source or an artificial neutral

EXTRANEOUS CONDUCTIVE PART (4)

A conductive part liable to transmit a potential including earth potential and not forming part of electrical installation for e.g. non insulated floor/walls, metal framework of buildings, metal conduits/pipe work (not part of electrical installation) used for water, gas, compressed air, etc.

BONDING CONDUCTOR (5)

A protective conductor providing equi -potential bonding.

MAIN EARTHING TERMINAL (6)

The terminal or a bar provided for a connection of protective conductors including equipotential bonding conductors and conductors for functional earthing

TEST LINK (7)

Allows an earth electrode resistance check.

Salient Points

- All enclosures, body, panel, steel structures, Substation fence etc. shall be grounded at several earth points and connected to ground bus. There should be two distinct connections for all equipment.

- The neutral point for each system like generator, transformer to be connected through its own two independent earth.

- Equipment framework and other non-current carrying parts to be earthed.

- All plug sockets shall be of 3-pin type and third pin shall be permanently earthed.

- All extraneous metallic framework not associated with equipment to be earthed.

- Earth connections should be visible and properly done.

- Testing of earth pits should be possible by means of usage of testing link.

- Size of earthing to be verified while adding any additional loading in system.

- Earthing system in a substation to be at a uniform potential and near zero or at absolute earth potential as possible.

- Earthing should provide low impedance path to fault currents to ensure prompt and consistent operation of protective devices during ground faults.

- Earth pits value should be less than 5 Ω and in substation will be less than 1 Ω.
- Drawing should be available for earthing scheme.

WIRING

Method of installation can be in non-metallic rigid PVC conduit if it is used concealed or metallic MS/GI conduit for surface/concealed or trunking or in ducting concealed in civil structure.

Number of wires to be laid are determined by standard and no way excess numbers to be installed.

Also, conduit is always having bends and hence wires installed in straight conduit is never to be considered but to consider allowed number of wires in bend conduit.

Given below is table which gives number of wires in rigid metallic conduit.

Number of Single core wires to be installed in conduit.

Sr. No.	Sq mm	20 mm	25 mm	32 mm	40 mm
1	1.5	5	10	14	-
2	2.5	5	8	12	-
3	4	3	8	10	-
4	6	2	5	8	-
5	10	-	4	7	-
6	16	-	-	3	6

As such approximately 20% space is only used in conduits by wires and not excessive wires to be installed inside.

The cross-sectional area of every wire shall be such that its current carrying capacity is as per load current without increasing temperature more than insulation. Following table will give approximate current carrying capacity in concealed wiring. This will differ from manufacturer to manufacturer and type / quality of insulation-

Sr. No.	Sq mm	Amp
1	1.5	16
2	2.5	20
3	4	26
4	6	33
5	10	45
6	16	60

The voltage drop between the origin of the installation and the fixed current-using equipment should not exceed 4 percent of the normal voltage of the supply.

Conductors of wires should be made out of electrolytic grade copper.

Insulation of wires should be such that it will not break below predetermined temperature which may cause leakage and then fault.

Voltage grade will be 1100 Volts.

Wires should have conformation standard of that country where it is going to be used and ISI mark in India.

Standard gives resistance of wires and hence wires must maintain the resistance. In case of increase in resistance

temperature of conductor and hence of wires for given current rating. Ultimately there will be insulation failure.

In wires there are various types and selection to be done accordingly like –

- FR having properties of flame retardant so that the propagation of flame is retarded. It has properties of limited oxygen index >29%, limited temperature index >250^0

- FRLSH having properties of flame retardant low smoke low halogen so that there is restriction of the spread of flames in fire situation and the smoke emitted by the burning of cable is considerably low compared to traditional cables. With this there is improved visibility for evacuation of trapped victims and facilitates firefighting operation. It has properties of limited oxygen index >29%, limited temperature index >250^0, Smoke Density (Light Absorption) < 60%, Acid Gas Generation < 20%.

- HRFR having properties of heat resistant flame retardant so that it retards the propagation of flame without compromising safety. It has properties of limited oxygen index >29%, limited temperature index >250^0

- HFFR Halogen free flame retardant as per IS 17048 having properties such as withstanding temperature range from -15°C to +90°C, insulation does not burn, melt and drip. Smoke is negligible, transparent and non-toxic. It is self-extinguishing and flame retardant.

Generally following sizes of wires are used for circuit and mains-

CIRCUIT AND MINIMUM COPPER WIRE SIZE

Sr. No.	Description	Sizing
1	Lighting	2x1.5 sq. mm +1x1.5 sq. mm
2	6 Amp socket outlet	2x2.5 sq. mm +1x1.5 sq. mm
3	16 Amp socket outlet	2x2.5 sq. mm +1x1.5 sq. mm
4	Water Heater <3KW	2x2.5 sq. mm +1x1.5 sq. mm
5	Water Heater >3KW	2x4 sq. mm +1x2.5 sq. mm
6	AC > 1.5 Tr	2x4 sq. mm +1x2.5 sq. mm
7	Appliance rated >3 kW<6 kW	2x6 sq. mm +1x4 sq. mm
	Meter to DB following minimum wiring to be used. In case of more load and increase length size to increase to compensate voltage drop.	
1	1 BHK	2x6 sq mm +1x4 sq mm
2	2 BHK	2x10 sq mm +1x6 sq mm
3	3 BHK	3x6 sq mm +2x4 sq mm
4	4 BHK	3x10 sq mm +2x6 sq mm

CABLES

Cables are widely used and basics are known to all. Hence theory part of the same is not being covered but design aspect and salient points related to 'electrical fire' are being highlighted. When cable is being thought what is being mainly discussed is current rating and type means PVC or XLPE and now some times selection out of FR PVC or FRLSH PVC or LSZH PVC.

PVC Cables advantages-

- Insulation is non-hygroscopic and unaffected by moisture.

- Complete protection against wide range of electrolytic and chemical corrosion.

- Fire resisting qualities are improved due to tough and resilient sheath.

- Ageing factor is improved.

- There is no compound inside which can migrate and hence installation in any direction is possible.

XLPE Cables advantages-

- Current Rating is high

- Service life is more

- Dielectric losses are less.

- XLPE is resistant to acidic and alkalis environmental influences.

- Short Circuit ratings are more and hence stresses, temperature in case of short circuit is enhanced.

- Manufacturing process is thermosetting so better handling properties, external effects and in manufacturing like in cooling.

- Internally it is cross-linking process so less stresses.

- Thermal resistivity of cross-linked material is low compared to thermoplastic material.

In case of PVC where one can have option of –

- Normal PVC

- HR PVC- Heat Resistant PVC

- FR PVC – Fire Retardant PVC

- FRLSH PVC – Flame retardant,
  Low smoke low halogen

- LSZH PVC - Zero Halogen low smoke PVC

Comparative statement of properties as given below-

Comparitive Properties of House Wires

Feature	Normal PVC Wire	Heat Resistant HR PVC	Fire Retardant FR - PVC	Flame Retardant Low Smoke FRLS	Zero Halogen Low Smoke
Insulation Material	PVC	PVC	Spl. PVC	Spl. PVC	Spl. Polymer
Insulation Property	Normal	Good	Good	Good	Very Good
Temperature Rating	70°C	85°C	70°C	70°C	85°C
Thermal Stability	Normal	Very Good	Good	Good	Very Good
Flame Retardancy	Good	Good	Very Good	Very Good	Excellent
Safety During Burning	Average	Average	Good	Good	Excellent
Requirement of Oxygen to Catch Fire (% in air)	→21	→ 21	→ 30	→ 30	→ 35
Temperature Required to Catch Fire (with 21% oxygen)	Room Temp.	Room Temp.	→ 250°C	→ 250°C.	→ 300°C
Visibility during Cable Burning (%)	←20	← 20	← 35	→ 40	→ 80
Release of Halogen Gas during Burning (% by weight)	←20	← 20	← 20	← 20	ZERO
Abrasion Resistance during Installation	Good	Good	Good	Good	Good

Photo from KEI

In case of cable selection along with catalogue current ratings, rating factors are also required to be checked. Because when condition is not ideal, current rating of cable reduces which is ignored by many. Hence following tables give rating factor which are widely used in industry. For this cables manufacturers catalogues can be also referred.

Rating factors related to variation in ambient air temperature			
Air Temperature in Degree.	30°	40°	50°
PVC	1.16	1	0.8
XLPE	1.11	1	0.88

Rating factors related to variation in ground temperature			
Air Temperature in Degree.	30°	40°	50°
PVC	1	0.87	0.71
XLPE	1	0.91	0.82

Rating factors related to variation in ground thermal resistivity of soil for multi core cables laid direct in ground.			
Thermal Res. in o C.Cm/W	150	200	300
Rating factors	1	0.9	0.76

Rating factors related to variation in depth of laying for 1.1 kv cables. For cross-sectional area of conductor 25 to 300 sq. mm				
Depth of laying (cm) >	75	90	120	150
Rating factors	1	0.98	0.96	0.94

Group Rating Factors for cables laid in ground				
No of cables in groups	Touching	S=15 cm	S=30 cm	S=45 cm
2	0.8	0.84	0.87	0.9
4	0.62	0.69	0.75	0.8
6	0.55	0.62	0.69	0.75
8	0.5	0.57	0.66	0.72
S=axial spacing of cables				

Cables laid in open rack in air				
No. of Racks	**No. of cables per rack**			
	1	2	3	6
1	1	0.98	0.96	0.93
2	1	0.95	0.93	0.9
3	1	0.94	0.92	0.89
6	1	0.93	0.9	0.87

No of cables in groups	No. of Tier	Touching	S=15CM	S=30CM	S=45CM
2	1	0.8	0.84	0.87	0.9
4	2	0.6	0.66	0.79	0.83
6	2	0.51	0.57	0.63	0.67
8	3	0.46	0.51	0.56	0.6

On basis of rating factors multiplication is done and final rating factor is calculated which is applied to current rating of cable and derate the current accordingly.

Further cables are selected on basis of Short circuit current rating having Relationship is $I^2t = k^2S^2$

Depending upon short circuit rating (I), fault clearing time of breaker(t) and material constant(k) area S is calculated.

Cables when are installed underground then standard depth will be as -

Voltage Grade	Depth of Laying
1.1 KV Cables	750 mm
3.3KV to 11 KV	900 mm
More than 11 KV	1050 mm

- Voltage drop is calculated and should not exceed 4 percent of the normal voltage of the supply.

- In case of motor load starting current to be seen and accordingly cable to be selected but again after applying the rating factor.

- Additionally standard selection criteria to be adopted as-

    - Voltage grade

    - Availability

    - Installation conditions such as bending radius if possible

    - Conductor if aluminium or copper

SWITCHGEARS

| MCCB | MCB | RCD | AFDD |

Photo from Schneider Electric

MCCB

MCCB means molded case circuit breaker is widely used in electrical distribution circuit. Mainly it gives over current, short circuit and earth fault protection. Standard for the same is IS/IEC 60947-2. Many a times usage is known and not only widely handled but used also. But still there are lot of gaps while selecting right technical specification of MCCB. Unless right specification is used functioning of MCCB will not happen which it should have happened. Hence following points will cover the salient points of selection-

Rated Current (In)- MCCB is having rated current with adjustable knob so that as per load current MCCB current can be adjusted. Above that current if load current increases it should trip. Generally, it starts from 63 Amp to 1600 Amp. As per present and future loading same can be selected considering protection to load and cables.

Frame Size – Manufacturers design the MCCB in few frame sizes and as per that size breaker can be rated to maximum rating. One frame can take care of many ratings of MCCB. Importance of frame size selection is for space planning in panels and future load addition. One frame can take care of 100 Amp to 200 Amp. So, in case load is going to increase than 200 Amp then higher frame size can be selected now. So that danger of overloading in future can be avoided. Also, other way round optimum sizing can be done by effective design for saving cost and space.

Rated working voltage (Ue)- Continuous voltage for which the MCCB is designed such as for three phase 400V or 415 V. For single phase it is 230V or 240V.

Rated Insulation voltage (Ui)- Maximum voltage which MCCB can sustain.

Rated impulse withstands voltage (Uimp)- Transient peak voltage the circuit-breaker can withstand from switching surges or lighting strikes imposed on the supply. Generally, it is 6 to 8 KV.

Frequency – Operating frequency if 50 or 60 Hz.

Poles- MCCB can be Single Pole (SP), Double Pole (DP), Triple Pole (TP), 3 Pole with Neutral (3P+N) and 4 Pole (4P).

Short circuit capacities –

Ultimate short-circuit breaking capacity (Icu) is highest fault current capacity which can clear maximum level of fault current without any damage to MCCB and can be re used. It is from 16 KA to 65 KA.

Service breaking capacity (Ics) is rated short-circuit current circuit breaker can break which is % of Icu. Since fault may come lower than ultimate breaking capacity.

Short circuit withstand capacity (Icw) is maximum short-circuit current which the circuit-breaker can withstand for defined period (0.5 or 1 or 3 s) without alteration of its characteristics.

MCCB Characteristics- MCCB operating characteristics are presented on time/current characteristic curves called as tripping curves. It shows the time required to trip at a given overcurrent.

When over current happens bi-metal conductor heats up and deflects causing breaker to trip. In case of short circuit which is 5 to 10 times of rated current, magnetic tripping happens and breaker trips instantaneously. Setting of the same can be done. In case of breakers having electronic trip one can adjust the overcurrent and short circuit rating as required.

Accessories – Few accessories which are to selected before ordering are - Shunt trip coil for remote tripping, Undervoltage coil for sensing under voltage, Earth fault sensing internally or externally, if motorized, Interlocks if required, auxiliary contacts, operating handle.

MCB

MCB means miniature circuit breaker is most widely used in electrical distribution circuit. Standard for the same is IS/IEC 60898-1. Mainly it gives over current, short circuit protection. Specification used for selecting MCB are-

Rated Current (In)- MCB is having fixed rated current starts from 0.1 Amp to 100 Amp.

Rated working voltage (Ue)- Continuous voltage for which the MCB is designed such as for three phase 400V or 415 V. For single phase it is 230V or 240V.

Rated impulse withstands voltage (Uimp)- Transient peak voltage the circuit-breaker can withstand from switching surges or lighting strikes imposed on the supply. Generally, it is 4 KV.

Frequency – Operating frequency if 50 or 60 Hz.

Poles- MCB can be Single Pole (SP), Double Pole (DP), Triple Pole (TP) and 4 Pole (4P).

Short circuit capacities –

Short-circuit breaking capacity is highest fault current capacity of 10 KA, which can clear maximum level of fault current without any damage to MCB and can be re used.

MCB Characteristics- MCB operating characteristics are presented on time/current characteristic curves called as tripping curves. It shows the time required to trip at a given overcurrent. When over current happens bi-metal conductor heats up and deflects causing breaker to trip. In case of short circuit, magnetic tripping happens and breaker trips instantaneously. In MCB there is differentiation of B, C, D Curves. In B curve instantaneous tripping happens between 3 to 5 times of rated current. In C curve instantaneous tripping happens between 5 to 10 times of rated current. In D curve instantaneous tripping happens between 10 to 20 times of rated current. Setting in case of MCB are electromagnetic only and not electronic hence are fixed.

RCD

RCD means Residual Current Device came in to product line to protect human being from electrical shock. Standard for the same is IS 12640-1, IEC 60947-2.

Effect of current on human body is given by graph as per standard IS/ IEC 60479-1

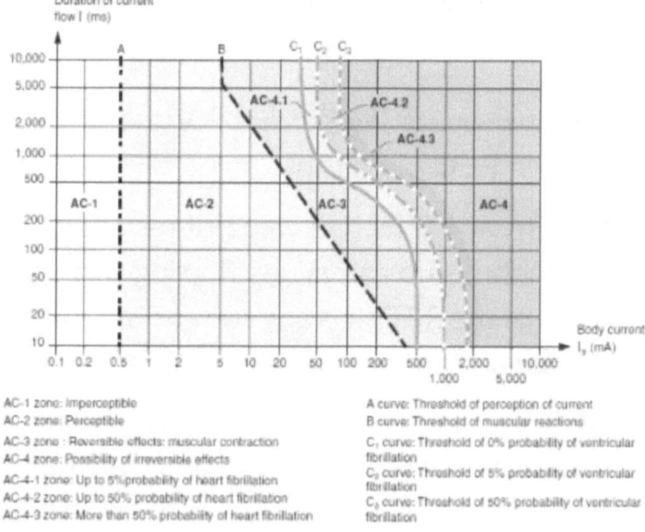

Currents from 10 to 30 mA are not fatal but their prolonged, presence causes muscular spasms, breath problems etc. currents above 30 mA may be fatal unless the person is quickly separated from the source Currents up to 500 mA will cause death if they pass for longer than 0.5 s; Currents above 500 mA are usually fatal even in short exposure times.

RCD works on the principle of comparing currents in phases and neutral passing through summation current transformer. In normal conditions the vectorial sum of current values equals to zero. When there is leakage due to insulation failure then part of the current starts to flow towards body. Hence there is imbalance condition in current transformer which generates current. This current activates relay and gives command to trip RCD.

Knowing this principle, the parameters for selection of RDC are-

- Rated residual operating current - Value of residual current when RCD must trip.

- Residual current - Effective value of resulting vector of instantaneous current values flowing through the main circuit of the RCD. It is any current value lower or equal or higher than rated residual operating current.

- Residual non-tripping current - Value of residual current at which the circuit breaker will not trip. Values are pre-set at manufacturing plant at 0.75 of rated residual operating current.

- Limit non-operation time – The non-operation time at which residual current device does not respond to residual current for time delay.

- The main parameter of a residual current device is rated residual operating current having values 30, 100, 300 mA. The main parameter of a residual current device is rated residual operating current having values 30, 100, 300 mA.

- Operating Temperature- In indoor installations, the range of ambient temperature from -5 °C to +40 °C is usually considered sufficient.

- Rated Voltage - Rated voltage of RCD the circuit breaker is designed for 110V/230V/400V.

- Frequency-50Hz

- Type- AC, A,B,F etc. Generally, AC type being used and all other are with DC component which is for particular usage.

Variants in RCD are RCCB and RCBO. RCCB is Residual Current operated Circuit Breakers without integral overcurrent protection. Whereas RCBO is Residual Current operated Circuit Breakers with integral Overcurrent protection.

AFDD

Arc Fault Detection device AFDD is new product development in electrical industry. Standard for the same is IEC 62606. This is used for detection of small arcs in wires and equipment.

Arc fault current is very small current which is developed due to insulation failure and arc is getting formed. This insulation failures could be due to piercing nail/ drill bit, insulation aging, rodents bite, forceful bending of wires, burning etc. After a period, this small current turns in to the big fault. When fault occurs on the same wire it is said as serial arc faults. Whereas when fault occurs between phase and neutral it is said as parallel arc faults. Such small arc develops high frequency noise. AFDD constantly monitor and analyses current and voltage waveforms and in case of any change happens which remains in circuit then solid state checks the change and treat as fault.

AFDD also works with RCD so that single device to give protection against over current, short circuit, earth fault and arc fault.

If comparison is done in MCB, RCD and AFDD then in short it can be said as-

MCB will trip at overcurrent of 1.45 times and short circuit of 5 times of rated current. RCD will trip at 30MA/100MA/300MA. Whereas AFDD will trip at all above conditions & at Arc of > 2 Amp, Over voltage >270 V.

Salient points in technical specification will be as below-

- Number of poles -2

- Tripping characteristics -2

- Rated current – As per manufacturer but generally 16 Amp

- Rated switching capacity - 10 KA

- Rated fault current – 30 mA

- Operating voltage – 230 V

LIGHTNING

Lightning is natural phenomena during thunderstorm. In this spark discharges take place from one cloud to another or between a cloud and the ground. The current in most cloud to ground flashes is negative flow from cloud to ground, ranging from 2000 Amp to about 200 000 Amp. Lightning can happen directly on structure, near structure, on electrical line or near to electrical line. But in all cases, it can be fatal giving damage to structure or equipment and most importantly human life.

Lightning protection is required or not is being decided by designers on basis probability of lightning which will struck

on building in a year calculation considering number of lightning flash density declared by standard and effective collection area of building. And then lightning protection system is being provided. So also, detailed design of lightning protection system to be done by designers. For information and verification following data is made available.

The lightning protections system comprises of - a) air terminations b) down conductors c) joints and bonds d) testing joints e) earth terminations sf) earth electrodes.

Air termination consist of vertical or horizontal conductors or combinations of both. All metallic projections, steel reinforcement in column beams, surface of the roof metallic coping, roof coverings, handrails, window and all metallic part of building should be considered for air termination network and to be bonded together to the earth making equipotential bonding. Material used for air termination, down conductor fixed connections is of aluminium, copper or galvanized MS steel of size 20 mm x 3 mm. Down conductor are finally connected to earthing pits made out of plate, GI pipe, GI flat or copper coated steel rod as per standard.

In case of building having rectangular roof, horizontal air termination consisting of conductor around roof is use along the periphery of a rectangular building. In case of structure having very small roof one air termination of 10 mm dia, 2 Mtr height is sufficient provided it covers area of building considering 45-degree cone from top. Similarly, multiple vertical air terminations are also used. But most of the cases horizontal conductor is used.

Generally, in case of tall reinforced building horizontal conductors on roof forming 10 m x 20 m network is used. Down conductors can be ascertained by periphery length divided by 20.

For building having height less than 60 Mtr side flashes of lightning are not considered. But for more than 60 Mtr upper portion of 20% of building will have air termination system.

Apart from direct strike there are surges which occurs up to two kilo meters due to lightning strike or even due to switching operations in large electrical systems. Due to these surge voltages, electronic components get damage permanently.

As such surge protective devices SPD are suggested to be provided in incoming line Class B+C, 40 KA at entry level and also at load side Class C, 20 KA for protection. For data lines Class D SPD to be use based on application.

Distribution Boards

Distribution Board (DB) term is very vast. Although it is mainly relating to small distribution boards comprising of like MCB, RCD, MCCB. But we can extend the definition to cover fire topic, from small DBs used in residence up to panels in residential complex/ commercial offices/ industries etc. Fire point of view all these distribution boards/panels are very important.

Main components into DBs/Panels are switchgears, fabrication, busbars, cables/wires compartment, earthing. Prima facie, it appears to be very simple and always discussion happens from few users that what is big deal in to manufacturing the same. Arguments always happens that why panels/DBs made by selected one or reputed one or OEM or their system integrators or only to be preferred. Hence one must be careful in selection of vendors. <u>Since one cannot ignore the fact that failures or fires do occur in DBs and panels.</u> This clearly indicates that one has to increase the yard stick of manufacturing to level of accepted as per 'standard'.

Technical specification of all is already well placed in tenders, standards and matured enough from theory perspective. In which construction features, painting, busbars, assembly, testing etc. are always well covered. And as such for detail specification one can refer those documents.

As such only salient points are covered below from fire perspective and other points to be taken from standard specifications. Let us understand those points-

A) Distribution boards-

   While manufacturing these OEM (original switchgear equipment manufacturer DB) are more preferred as they will be duly type tested and capturing minute features which helps for safe installation. Salient points can be –

   1) It shall be manufactured as per latest standard which IEC 61439.

   2) Fabrication will be with high quality Cold rolled cold annealed sheets.

3) It will have basic structure suitable for flush or surface mounting, independent door, shield, removable gland plate and pan plate with DIN rails. Collar will be provided in case of conceal type for plaster guide. This will help for easy installation and periodic maintenance.

4) DB shall be with IK08 for single door and IK09 for double door, protection against mechanical and IP43 or IP54 protection range as required.

5) Doors will have gasket for enhanced IP protection.

6) It will have fully insulated Busbars and removable shrouded Neutral links for protection to the user in maintenance activity along with earth busbar.

7) The Copper Busbars shall be having capacity of 100 A minimum for MCB DBs and for MCB DBs same will be as per rating of incoming for MCCB.

8) It will have adjustable neutral bar position for ease of wiring mounted on insulators.

9) Wherever required, busbars will have fishbone assembly.

10) Plastic caps, plugs for covering unused terminals, openings will be used.

11) Sufficient space to be provided for wiring. Wire inside will be FRLSH, tied by ties which shall come along with DB.

12) Neutral busbars will be color coded for easy identifications for per phase isolation.

13) It should have integrated door earthing system offering additional safety to the user with earth marking.

14) DB identification labels will be provided on plate.

15) DB shall be supplied preferably with masking sheet to prevent cement entering the box during plastering.

16) Lugs will be used on wire terminals.

17) Clear marking, indication and mounting instructions shall be provided along with DB.

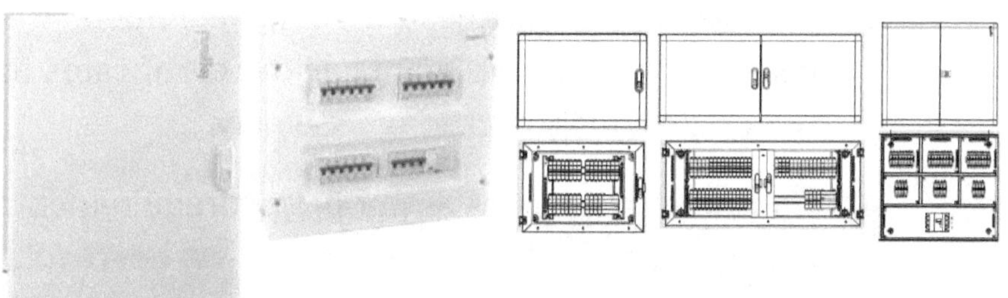

Photo from Legrand

B) Panels

While manufacturing Panels OEM or their system integrator are more preferred as their panels will be duly type tested and capturing minute features which will help for safe installation. Panels manufactured should be conforming to the latest standard which is presently IEC 61439. In this apart from specification, roles and responsibilities of supplier are clearly defined for verification by various parties involved. Same must be followed as given below-

1) Original Manufacturer will make the original design and associated verification of an assembly system and will be responsible for the "design verifications".

2) Panel Builder will make panels as per instructions of the original manufacturer and will be responsible for routine verifications on each panel produced, according to the standard. If there is any deviations from the instructions of the original manufacturer then they have to repeat the design verifications.

3) In both cases certification body conducts the verification tests and grants all the certificates to the original manufacturing assembly.

4) Consultants specifies as per project requirements and checks if all requirements have been fully carried out by panel manufacturer.

5) End user or client can always request a certified LV switchboard with systematically carried out routine verifications, testing etc.

The 10 main functions of standard are-

1) Voltage stress withstand capability- Insulation to withstand long-term voltages, transient and temporary over voltages guaranteed through

clearances, creepage distances, and solid insulation.

This is checked by: Measurement of clearances and creepage distances, Power frequency dielectric test, Impulse with stand voltage test

2) Current-carrying capability- Protect against burns by limiting excessive temperatures which is being checked by single circuit is continuously loaded to its rated current and any circuit is continuously loaded to its rated current multiplied by its rated diversity factor.

   This is checked by: Temperature rise tests

3) Short-circuit withstand capability- Withstand short-circuit through protection devices, short-circuit coordination and capability to withstand the stresses resulting from short-circuit currents in all conductors.

   This is checked by: Short-circuit tests

4) Protection against electric shock- Hazardous live parts are not accessible and accessible conductive parts are not hazardous for life.

   This is checked by: IP XXB test and verification of insulating materials, mechanical operation tests, verification of dielectric properties, measurement of the resistance between each exposed conductive part and the PE terminal, Short-circuit strength of the protection circuit

5) Protection against fire or explosion hazard- Protect persons against the fire hazard by resistance to internal glowing faulty elements through selection of materials and design provisions.

   This is checked by: Glow wire test

6) Maintenance and modification capability- Capability to preserve continuity of supply without impairing safety during assembly maintenance or modification through basic and fault protection and optional removable parts.

   This is checked by: IP tests, Mechanical operation tests

7) Electro-Magnetic compatibility- Properly function and avoid generation of EMC disturbances through incorporation of electronic devices complying with the relevant EMC standard and their correct installation.

   This is checked by: EMC tests according to product standards

8) Capability to operate the electrical installation- Properly function, according to the electrical diagram and the specifications, the specified operating facilities through accessibility and identification.

This is checked by: Inspection, Impulse with stand voltage test of isolating distance for withdrawable units

9) Capability to be installed on site- This is being done by withstand handling, transport, storage, and installation constraints and capable to construct and connect through selection or design of the enclosure and the external terminals and by provisions and documentation.

This is checked by: Lifting test

10) Protection of the Assembly against environmental conditions- Protect the assembly against mechanical and atmospheric conditions through selection of materials and design provisions.

This is checked by: IP test, IK test, Corrosion test, UV test for outdoor panels

Idea behind all this is to make panel designed properly at drawing stage then order the components accordingly and frame up the design. Same design to be verified by testing in accredited third-party laboratory. And same design will be implemented in further manufacturing. In case design is transferred to system integrator then they will also follow and repeat the design. Then if changes are taken place or in case one makes deviation then design will be retested. With these all-loose connections, are tied together and output

Panel is full proof tested and ready to use safely. It is therefore suggested to have panels conforming to latest standard.

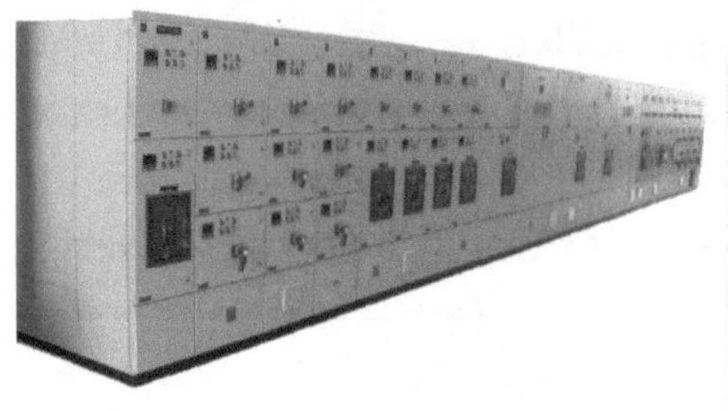

Photo from Arrow Engineers

Photo from Schneider Electric

Salient points will be-

Busbars

a) Electrolytic grade aluminium busbar having conductivity of 56% and copper busbar with high conductivity of 99 to 100% to be used for bus bars. In any case commercial grade should not be used. All the busbars should be tested for purity with conductivity meter at factory to make sure the purity is maintained. Then one can use busbar as per predetermined current density.

b) Sufficient space in busbar chamber with proper clearance of busbar to avoid overheating of busbar

c) Busbar configuration should be type tested.

d) Busbar arrangement should be such that proximity and skin effect are minimized so that heat losses are reduced.

e)   Clearances between busbars should be sufficient and not as per rule of 25 mm between Ph to Ph.

f)   The contact area must be at least 5 times the cross-section of the bar.

g)   Busbar inside the enclosure must be totally encapsulated.

h)   It is advisable to place the neutral bar at the front of the busbar for additional safety, easy connection of circuits supplied between phases and neutral, easier identification of the neutral earthing system if applicable and reduction of the radiated magnetic field.

i)   Nut bolts to be fixed in manner of bolt, plain washer, busbar, plane washer, spring washer and then nut. Contact should be tight but the contact pressure should not be too high, else deformation of busbars will happen and hence always use torque wrench. With which limit of elasticity of the bar will not be exceeded.

j)   After providing tightening torque, the marking should be done with some paint

k)   Use Zinc passivated hardware of tensile strength Grade 8.8 for busbar

l)   Busbar Supports used Glass Filled SMC which should pass glow wire test.

Fabrication

a) Design should be generally modular and with standard architecture, made out of 14/16 SWG CRCA Sheet for body and 10 SWG for Gland plate for ACB Feeder.

b) It should be modular bolted design which can add to quality, finish and easy assembly. Added advantage with bolted design is that to change feeder compartment size is possible. But in this case body should be fully tested for all test including arc flash, short circuit and must be proven design. Compartmentalization is needed as per forms mentioned. The modularity of the switchboards ensures that one can modify or upgrade them with ease to adapt for changing processes and increase performance.

c) Standard defines the separations inside an assembly according to 4 types of form, each form being divided into two groups, "a" and "b". Form 4b is desirable in which separation of busbars from functional units and separation of all the functional units from each other including terminals for external conductors. Terminals for external conductors are not in the same compartment as the functional unit but in separate individual compartments. Safety protection with terminal block covers is advisable with internal separation of 4b.

d) Degrees of protection can be minimum IP54 for dusty and damp environments and will vary as per environmental condition.

e) Live parts are protected by screens guaranteeing an IP20 degree of protection.

f) Switchboards should incorporate the best safety and protection system so that on-load operation becomes safer.

g) When the draw out units are used then same should allow to work safely outside the switchboard while it is energized, therefore, ensuring continuity of service throughout the process. The switchboard is designed with free slots to add new functions or extra motor feeders subsequently and urgently. The withdrawable drawers to have three positions as connection, disconnection and testing. Tap off equipped with shutters prevent access to live parts when the functional unit is removed out from the column.

h) Ample space in Cable chamber for easy termination of cables at site

i) Door hinges should be preferably concealed type to give additional protection against unauthorized access to feeders.

j) Door design should be such that there will not be accidental opening.

k) Design should be such that operator should not in contact with live parts.

l) Construction should be such that less maintenance in opening, maintaining the panel is possible.

m) Doors should open at 90 deg for getting full access to compartment.

n) Cable alleys should not be restricted to 300 mm because at times aluminium cables are used and therefore heat dissipation happens properly and cable derating reduces.

o) Fire Retardant type Gasket to be used removable/openable parts.

Earthing

a) Earthing busbars should running through out panel and earth of each door and all structural members, with which stray voltage on door is absent.

b) The protective conductor collector to be marked with the earth symbol on the chassis or the main structure. It has a terminal for connecting the protective conductor of the power supply. Exposed conductive parts must be electrically connected to one another so that no dangerous voltage can arise between exposed conductive parts that are accessible simultaneously. This continuity can be obtained by construction or by using equipotential link conductors.

c) The panel must be built so that no voltage can be transmitted from the inside to the outside.

d) The symbol must be visible from the outside.

Other Points

a) Panel to be tested for Internal Arc at 50/65KA for 0.5sec for feeder compartment, HBBC VBBC busbar compartment & cable alley compartment

b) Panels to be certified for Seismic test for zone 5 for taking care of earthquakes and high vibration resistant.

c) To minimize the induction created in magnetic loops, it is always advisable to have all the live conductors and neutral in the same metal frames to avoid any creation of fields.

d) To minimize the induction created in magnetic loops, it is always advisable to have all the live conductors and neutral in the same metal frames to avoid any creation of fields.

e) While designing the panel environmental impacts must be calculated and to ensure that components used will have no impact on environment.

f) End of life optimization to be done so as to decrease the amount of waste and allow recovery of the product components and materials.

g) Verification of temperature rise limits. Busbars, connections, functional units have been tested in order to avoid connection damage, reduction of insulation performance, risk of burn and faulty operation of devices.

h)   Forced ventilation for environments with ambient temperatures hotter than 45°C or for devices with considerable heat loss.

i)   Panel painting will be with almost eleven tank process like degreasing, de rusting, chemical activation, phosphating, passivation, oven drying etc. and finally with polyester powder coating.

j)   Insulating materials used should be tested to withstand the electrical, mechanical and thermal stresses in condition of operation, during ageing and will be fire resistant.

SWITCH SOCKET

1)   Switches, sockets, switch socket combination are very widely used item everywhere. However, the same item can become very vulnerable when issue of fire comes. Because this small item is causing fire in many cases. So, one has to be very careful from this.

2)   Any loose connection in terminal will start sparking and will become starting point of fire.

3)   It comprises of GI box along with modular metal base, top cover plate, modular switches and cover flushed to wall.

4)   Base plate is normally metal with Antirust coating and with adequate mechanical strength. On this modular switch-Sockets are mounted.

5) Preferred materials used for tip contact should be Silver-Cadmium, rocker / plunger should be polycarbonate, screws should be steel with zinc plating and cover polypropylene FR type. Overall fire-resistant materials are preferrable.

6) Assembly to be made out of and self-extinguishing temperature which is as high as of 850ºC for 30 s.

7) Terminal arrangement should be tunnel type and angular so that there will hold to the location where it is terminated.

8) In some cases, Z plunger is used instead straight so that there is full contact separation.

9) Shutter arrangement for child proof / single pin insertion should be must in any socket.

10) One must ensure that it should have ISI mark.

Few latest likely amendments in BIS should be honored while selecting the materials-

a) Only three pin design is approved and 5 Pin design is not approved.

b) Oval shape design is not approved only for 6/16 A except for 2 pin it is permissible.

c) New construction for socket contacts, pins of sockets and screwless terminals is approved.

d) New test requirements for zero strain on non-rewireable plugs/sockets are added.

e) International socket is removed and circular pins sockets to be used.

f) 10 A Plug is deleted and instead of that 16 A Plug will be used.

Connectors and loose connection

Whenever single wire connection is there then it is advised to use lugs crimped to wire. With which all strands are bound together. In absence of lugs after tightening strands spread and contact area reduces. Hence essentially wires strands to be crimped.

Sometimes there is requirement of multiple connections to one unarmoured cable for connecting multiple power outlets on one circuit. At that time there is no option but to make a joint in wires or to have multiple wires connected in one socket as shown in following figure. In such case likely loose connections may happen causing heating and then spark and fire as shown below-

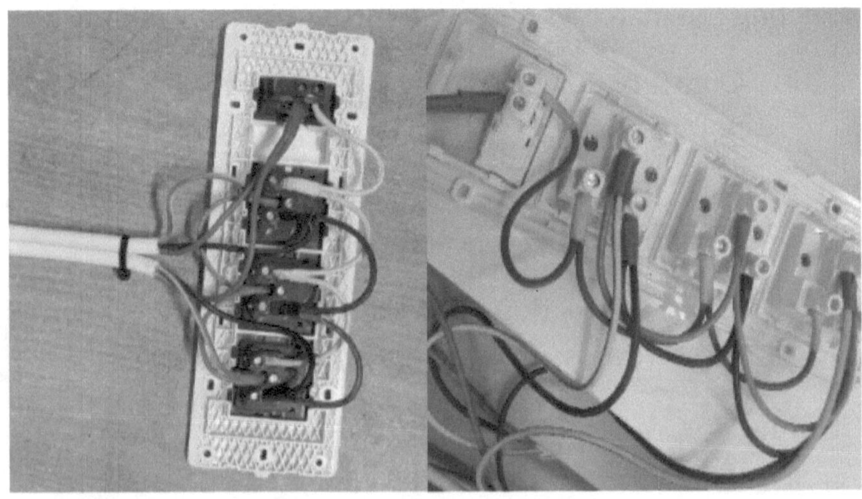

Instead, if customized connectors are used for power, then loose connection issue is not there, human error is avoided, in socket only one wire is connected and not multiple wires. Entire system becomes perfectly connected, plug and play and making ease in maintenance. This can be seen following photo. One can get these connectors as per international Standard and are made out of self-fire extinguishing engineering plastic housing which makes safe installation. Strands of wires to be crimped properly, put in connectors and tight the same and connect connectors properly.

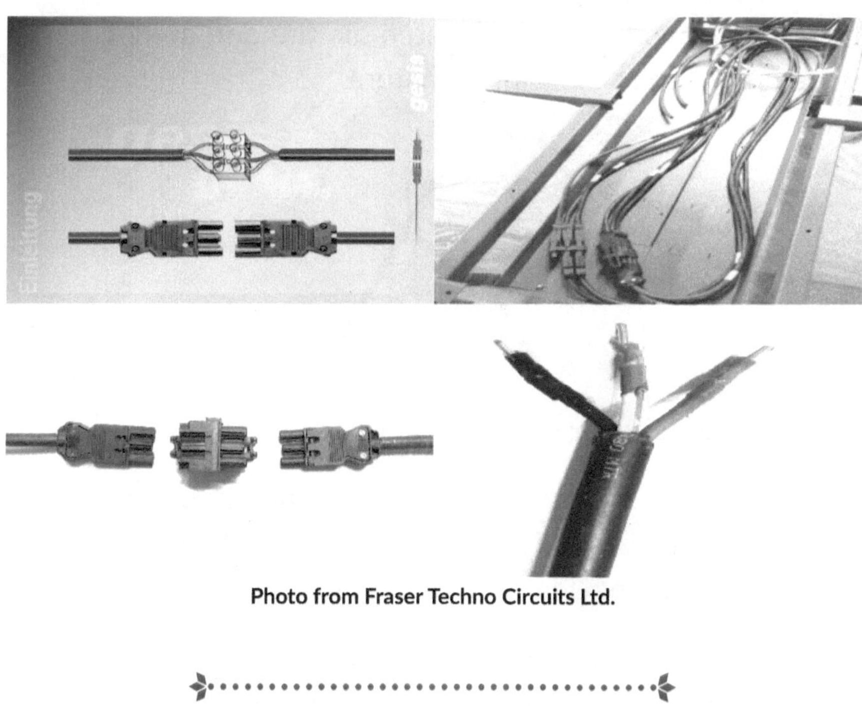

Photo from Fraser Techno Circuits Ltd.

FIRE THEORY

Fire was present on earth right from prehistoric time and early ages. As the progress of human being happened and further progress happened then it became part of all religions. In ancient Greek philosophy and science Fire is one of the four classical elements in earth, water, air, fire and aether, which were proposed to explain the nature.

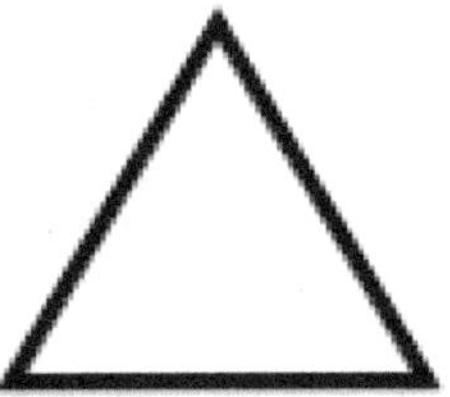

Image by Kwamikagami CC BY-SA 4.0 via Wikimedia Commons

In alchemy, which is tradition practiced throughout in Europe, Africa and Asia to purify, mature, perfect certain objects into gold. They were having symbol of upward-pointing triangle.

Agni is a Hindu and Vedic deity. The word agni is Sanskrit is for fire. Agni is supposed to be the god of fire and the acceptor of sacrifices. The sacrifices made to Agni go to the deities because Agni is a messenger from and to the other gods.

From mythological beginnings, the triangle, pyramid, or delta symbol "Δ" as shorthand representation of heat, fire, or light soon became the dominant symbol in science and particularly in chemistry.

That must be the starting of Fire Triangle. To explain Fire the Fire triangle is being used.

1. **The fire triangle.**

    The original Fire Triangle was developed as a simple way of understanding the factors of fire. The **fire triangle** or **combustion triangle** is a simple equilateral triangle model for understanding the necessary ingredients for most fires.

    The triangle illustrates the three elements a fire needs to ignite:

    - Heat,

    - Fuel, and

    - Oxygen.

    A fire naturally occurs when these elements are present and combined in the right mixture. And a fire can be prevented or extinguished by removing any one of the elements in the fire triangle.

    For example, covering a fire with a fire blanket removes the "oxygen" part of the triangle and can extinguish a fire.

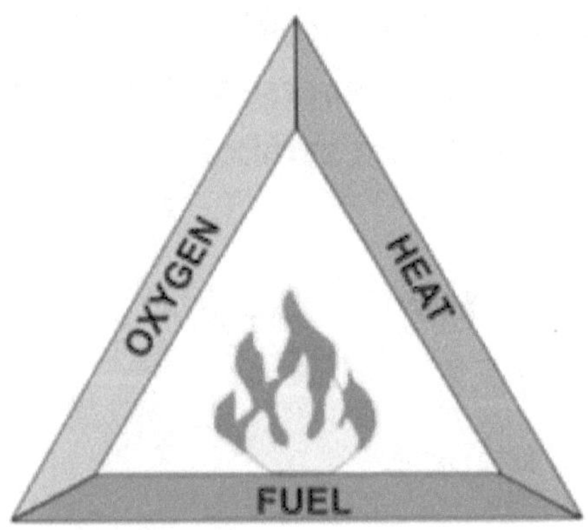

Image by Gustavb CC BY-SA 3.0 via Wikimedia Commons

HEAT is the usual way that to start fires. When heat is applied to some fuel then it ignites. There is always oxygen in our atmosphere, so we have the three necessary factors present. All primitive fire-making methods uses heat to start fire.

FUEL is what we add the heat to in order to start the fire. Once the fire is started, the heat from flames sustains the fire, and causes more fuel to ignite and burn.

OXYGEN is needed to cause combustion to occur. Removing oxygen is the usual way that we put out fires, such as with water or covering them with earth or snow. This has the effect of smothering the fire, cutting off the supply of oxygen. Without oxygen the fire dies.

2. **Fire Tetrahedron**

There is modification that happens in Fire Triangle to become Fire Tetrahedron.

The fire tetrahedron represents the addition of a component, to the three already present in the fire triangle is the chemical chain reaction. Once a fire has started, the resulting exothermic chain reaction sustains the fire and allows it to continue until or unless at least one of the elements of the fire is blocked. Foam can be used to deny the fire the oxygen it needs. Water can be used to lower the temperature of the fuel below the ignition point or to remove or disperse the fuel. Inert or any fire suppression gas can be used to remove free radicals and create a barrier of inert gas in a direct attack on the chemical reaction responsible for the fire.

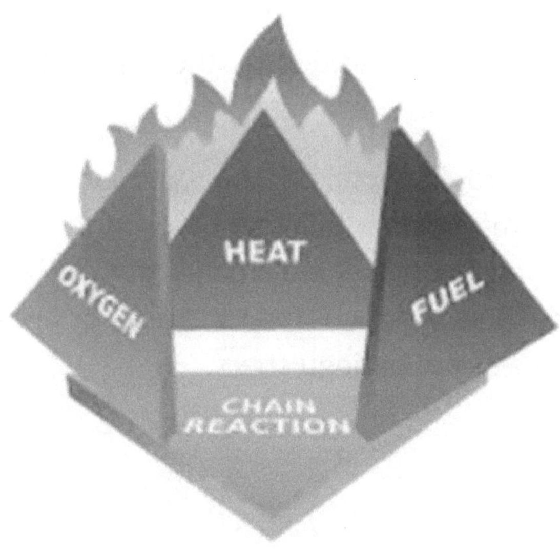

Combustion is the chemical reaction that feeds a fire more heat and allows it to continue. When the fire involves burning metals like lithium, magnesium, titanium etc. it becomes even more important to consider the energy release. The metals react faster with water than with oxygen and thereby more energy is released. Putting water on such a fire results in the fire getting hotter or even exploding. Carbon dioxide extinguishers are ineffective against certain metals such as titanium. Therefore, inert agents (e.g. dry sand) must be used to break the chain reaction of metallic combustion.

Fire class

All the fires are not same because the combustion materials or fuel which burn are different such as it could be paper, wood, oil, electrical cables etc. As such Fire classes are developed and Fire class is a term used to denote the type of fire, in relation to the combustion materials which would have ignited.

Fire Classes are being decided by the separate standards in the separate nations.

This has onward impacts on the type of suppression or extinguishing materials which can be used to douse the fire. Accordingly, Fire extinguishers are designed to tackle specific types of fire. There are six different classes of fire and several different types of fire extinguishers.

Image	Description	Class	Suitable suppression
	Combustible materials (wood, paper, fabric, cloth, plastics, rubber, coal, carbon-based compounds etc)	A	Most suppression techniques
	Flammable liquids such as grease, gasoline, oil, paint, thinners, kerosene, alcohol, etc	B	Inhibiting chemical chain reaction, such as water mist dry chemical or Halon
	Flammable gases such as LPG, Butane, Acetylene, Hydrogen, natural gas, Methane, etc	C	Inhibiting chemical chain reaction, such as dry chemical or Halon
	Flammable metals like magnesium, aluminium, sodium, potassium etc	D	Specialist suppression required or dry powder
	Electrical fire electrically energized (computers, switchboards, power-boards, cables etc)	E	As ordinary combustibles, but conductive agents like water not to be used
	Involving cooking oils and fats (most commonly occur in industrial kitchens)	K	Suppression by removal of oxygen or water mist

Class A: Ordinary combustibles

Class A fires can be denoted where combustibles materials are ordinary such as wood, paper, fabric, plastic and most kinds of trash. These fires follow the fire triangle. As such remove any one side of triangle and fire will get collapsed.

Such as covering a fire with a fire blanket removes the "oxygen" part of the triangle, use of water removes "heat" part of the triangle, removal of flammable materials or isolating flammable materials which is under fire from other nearby flammable materials removes "fuel" part of the triangle and one can extinguish a fire

Class B: Flammable liquid combustibles

Class B fires whose fuel is flammable or combustible liquid. These fires follow the fire tetrahedron (heat, fuel, oxygen & chemical reaction) and in which the fuel is a flammable liquid such as diesel, petrol, gasoline, or any other flammable liquid. Once fire starts, exothermic chain reaction starts and due to which the fire is sustain and allows fire to continue, until or unless at least one of the elements of the fire is blocked. Combustion is the chemical reaction that feeds a fire more heat and allows it to continue. Being difference between Class A and is this is chain reaction which makes fire to continue and increase.

One of the most effective way to extinguish a liquid fueled fire is by inhibiting the chemical chain reaction of the fire, which is done by dry chemical and Halon extinguishing agents (earlier it was used) or smothering with CO2 or for use of foam is also effective. Halon has fallen out of favor in recent times because it is an ozone-depleting material; the Montreal Protocol declares that Halon should no longer be used. Chemicals, gas such as CO2, Inert gas are now the recommended suppressant. In case of flammable liquid solid stream of water should never be used to extinguish this type because it can cause the fuel to scatter, spreading the flames.

Class C: Flammable gas

Class C fires same as above except whose fuel is flammable or combustible gas. Hence extinguishing element is different. These fires follow the fire tetrahedron (heat, fuel, oxygen & chemical reaction) and in which the fuel is a flammable gas such as propane, butane, methane or any other flammable gas. Dangerous thing is that gases travel significant distance and if they come in contact with ignition source then fire can rapidly propagate and spread. Also, it can cause explosions because some gases can erupt violently. Powder fire extinguishers can be used for flammable gases as powder within extinguisher will act as suppressing agent smothering oxygen within fire and stopping fire from scattering and spreading.

Class D: Combustible Metal

Class D fires whose fuel is combustible metals such as magnesium, potassium, titanium, zirconium etc. In case of some combustible metals fire risks sometimes reduces because they have the ability to conduct heat away from hot spots so efficiently that the heat of combustion reduces. This also means that it will require a lot of heat to ignite a mass of combustible metal. In some exceptions of the metals that burn in contact with air or water (for example, sodium), masses of combustible metals do not represent unusual fire risks because they have the ability to conduct heat away from hot spots so efficiently that the heat of combustion cannot be maintained—this means that it will require a lot of heat to ignite a mass of combustible metal. Generally, metal fire risks exist when sawdust, machine shavings and other metal 'fines' are present. These fires can be ignited by the same types of ignition sources that would start other common fires. Water and other common firefighting materials can excite metal fires and make them worse. The metal fires be fought with "dry powder" extinguishing agents. Dry powder agents work by smothering and heat absorption. The most common of these agents are sodium chloride granules and graphite powder. In recent years powdered copper has also come into use. Metal fires can pose more hazard because people are often not aware of the characteristics of these fires and are not properly prepared to fight them. Therefore even a small metal fire can spread and become a larger fire in the surrounding ordinary combustible materials.

Class E: Electrical

Electrical fires are fires involving potentially energized electrical equipment. This fire may be caused by short-circuiting machinery or overloaded electrical cables. In these fires water or any other conductive agents cannot be used by any one including firefighters. As electricity, may be conducted from the fire through water to the body and then earth and this may cause electrocution and deaths.

Electrical fire may be fought in the same way as an ordinary combustible fire except water, foam and other conductive agents are not to be used. The fire can be fought with any extinguishing agent rated for electrical fire. Carbon dioxide CO2, NOVEC 1230, FM-200 and dry chemical powder extinguishers such as PKP dry-chemical fire suppression agent and even baking soda are especially suited to extinguishing this sort of fire. PKP should be a last resort solution to extinguishing the fire due to its corrosive tendencies. Once electricity is shut off to the equipment involved, it will generally become an ordinary combustible fire. In some countries "Electrical Fires" are no longer a class of fire as electricity cannot burn but the items around the electrical sources may burn. By turning the electrical source off, the fire can be fought by one of the other class of fire extinguishers.

Class K: Cooking oils and fats

Fires that involve cooking oils or fats are designated "Class K" fire. These fires involve unsaturated cooking oils in well-insulated cooking appliances located in commercial kitchens. Such fires are technically a subclass of the flammable liquid/gas class but the special characteristics of these types of fires is the higher flash point which is considered important to recognize separately. Water mist can be used to extinguish such fires. Appropriate fire extinguishers may also have hoods over them that help extinguish the fire.

Overview of Fire, type of fires, classes etc. is seen above. Now let us move to various methods of fire detection and supporting equipments from life safety prospective.

LIFE SAFETY

Fire Detection

In case of fire there should be automatic fire detection system should be in place, so that fire incidence will be detected. This will give alarm to occupants to evacuate the place and call fire brigade. For which fire alarm system is used.

A fire alarm system comprises of intelligent fire alarm panel connected to number of devices working together to detect and warn people when smoke, fire, carbon monoxide etc. are present.

The fire alarm system comprises-

1. Fire Alarm Panel

    Fire alarm panel mainly three types. One is conventional type, second is addressable and third is wireless type.

    a) In case of Conventional fire alarm system, area is subdivided in zones and detectors are placed in zones. In case of fire occupant will come to know in which zone fire happened. Drawback of this system is exact location is not known but zone is understood.

b) In case of Addressable fire alarm system, address is given to each and every detector and device. And connection is done between fire alarm panel to each and every detector and device. Hence in case of fire information from specific detector is given to panel. As such exact location of fire is obtained to occupants. Predetermined number of detectors and devices are connected in each loop. Such loops can be multiplied. Hence it is widely used in offices and complex. Due to intelligent system and exact location of fire, smoke, heat is known same is being used. It has many advantages like sending messages to predetermined number. It can be seen on IBMS. It can give evacuation notice on public address system. Because of multiple advantages Addressable fire alarm system is widely used.

c) Wireless fire alarm system is wireless, battery power operated, radio linked system. It is expensive, flexible, quick to install where conventional cabling is not allowed. But this system is less adopted.

FIRE ALARM PANEL

a) Location of main panel will be near to main entrance. Repeater panels can be placed on floors or other entrances or in control room. There should be clear visibility near panel. Also, noise level should be low so that alarm will be audible. Main as well as repeater panels will be interconnected by network cables.

b) Main supply to panel will be from reliable source of supply such as mains and UPS with battery backup of 30 min at full load. MCCB supplying power should be painted as 'FIRE ALARM — DO NOT SWITCH OFF' preferably in enclosed box. Rating of the same will be sufficient to cater the load.

c) Cabling will be preferably with red color FRLS armoured shielded cable. Now a days cable having property of fire survival is also being used. Method of installation should be having return loop. Fire cables to be installed away from other power cables to avoid electromagnetic interference. Minimum size should be 2x1.5 sq mm copper conductor cable. In case unarmoured cable is used then it should be drawn through metallic conduit.

d) Appropriate glands shall be provided where the cable enters the junction box on which detectors are mounted.

e) Cables connected to detectors shall be given 'S' loop on both the sides of the detectors, sounders, panels etc. which shall be properly clamped to the ceiling.

f) Fire panel should be of modular expandable loop design and should support maximum 198 detectors and 198 control/monitor modules or each loop with 250 devices and modules in any combination or 150 detectors and150 devices or any other combination. This combination will change as per makes.

g) Panel should have facility to carry out programming conveniently and with easy operations. It should have special features such as cross zoning, day/night mode, selective control operations, etc. through external PC.

h) System parameters and control panel display and operation functions are conveniently programmed from the onboard LCD and keypad.

i) Panel should have auxiliary outputs and relays to be used for third party integration. There should be additional communications ports for Dialer, PC Interface, Voice Evacuation System, Remote

j) Annunciators, serial ports with connections for optional RS485 or RS232 plug-in communication cards or USB port.

k) Essentially panel should have in built battery charger for charging the sealed maintenance free 24 V batteries.

l) There should be event logger to know history.

m) Panel will have LED type general panel status indicators- Fire, Fault, Acknowledge, Disable test, Sounder fault/disable, Delayed mode, Relays disabled, Earth fault, System fault, alarm silenced, power supply fault etc.

n) The system can be used in environmental conditions -10% to 93% RH and $\leq$10 to $\geq$50 degree C ambient temperature.

o) Entire system can be approved as per UL, EN, FM, VDS etc. depending upon organizations.

2. Addressable Photoelectric optical type detectors are used for detection of smoke in premises. Heat detectors to be provided in cooking areas. Multi sensor detectors with smoke and heat detection facility are used for critical areas.

In case of smoke detectors there are two types of detectors. One is ionization type and other is optical type detectors.

In ionization type there is small amount of radioactive material between two electrically charged plates. This material ionizes the air and causes current to flow between the plates. When smoke enters the detector, it disrupts the flow of ions, thus reducing the flow of current and activating the alarm.

In optical type optical type smoke detector light source is aim into a sensor at an angle. When smoke enters the chamber, light is getting reflected away from the light sensor and which gives the alarm.

Due to presence of radioactive material, optical type detector is widely used. When smoke density exceeds the level smoke detector sends the signal to fire alarm panel and in case of increase of heat, heat detector gives the signal to panel. Then fire alarm panel gives alarm that there is fire.

Photoelectric
Detector

(All Images of detection and Fire panels are from MIRCOM)

3. Beam Detectors are used for detection of fire for heights more than 10 meters. It comprises a transmitter and receiver in a single enclosure and is usually installed between 19 inches and 24 inches below the ceiling. The transmitter emits an invisible infrared light beam that is reflected via a prism mounted directly opposite and with a clear line of sight. The reflected infrared light is detected by the receiver and analysed the signal. When Smoke arises then it reduces infrared light proportionally as per the density of the smoke which is in the beam path. The detector analyses this attenuation or obscuration of light and acts accordingly.

4. Manual Call Points are used to give signal manually to fire alarm panel. If one observes the fire then one can press break the glass and press the push button. This will give signal to fire alarm panel and sounder will get activated. They are placed at entrance, exits and in side premises as per standard.

Intelligent
Manual Station

5. Hooters with flashers used for audio and visual indications at predesigned locations. The hooters provide audible alarm. These hooters are connected to Fire alarm panel and receive power directly from the loop. There is provision for volume settings by means of switch. Also one can select tones of the sound. Generally sound output is 100 dB.

Hooter/Strobe

6. Response Indicators will be used for knowing conditions of detectors located above ceiling void or below false floor which are not visible. In case of fire and detector is activated then being response indicator connected to it, same will blink light and will indicate location of detector which has activated due to fire.

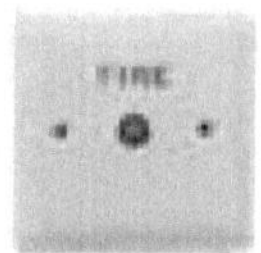

7. Isolation module to be given as per requirements for loop continuity. Generally, after 18/20 detectors the isolator is provided. If there is cable cut prior to isolator module then through isolator module and cable which is after isolator module system works. Hence provision of such modules are given.

8. Monitor / Control modules for integration to be considered as per requirement.

9. Fire Alarm System is integrated with other systems for smooth operations of other systems in case of fire –

   • Access control doors to open

   • AC system to off

- Pressurization Fans to start

- Basement Ventilation to start in fire mode

- Public address system to activate to give evacuation message

- Lifts to come at ground floor

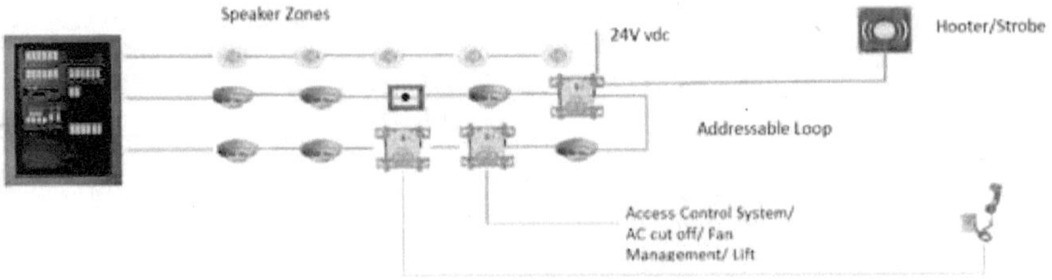

Conventional Fire Alarm system architecture

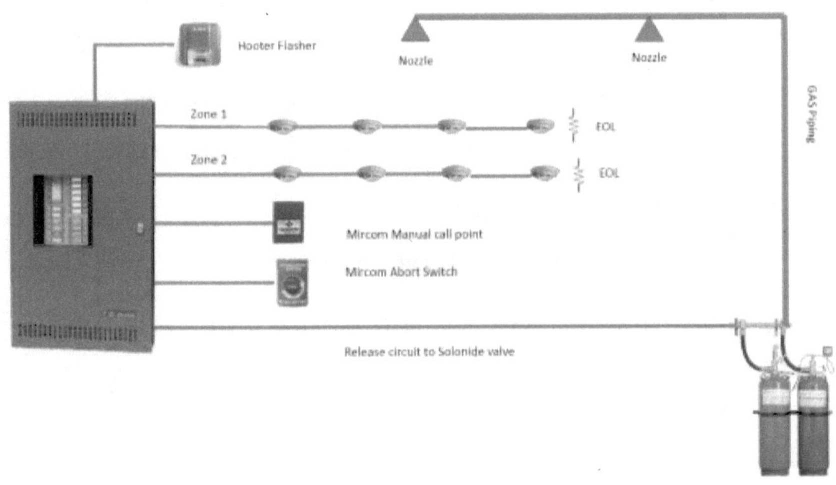

Addressable Fire Alarm system architecture

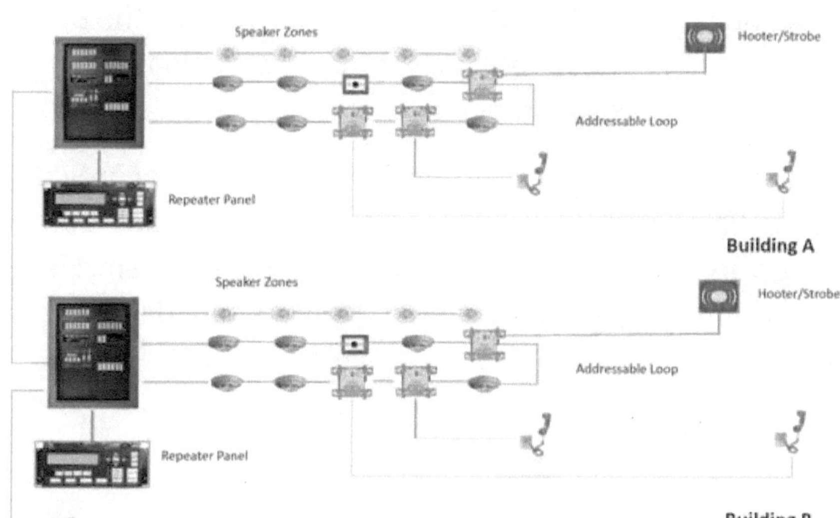

VESDA BASED EARLY FIRE DETECTION SYSTEM

In critical rooms early warning is required in case of fire. This detection is achieved using Very Early Smoke Detection Apparatus also called as VESDA. Also called as Very Early Warning Aspirating Smoke Detection solution.

The VESDA system consists of Air sampling tubes and Detector Assembly. The detector assembly consists of a laser detection chamber, high efficiency aspirator, monitored dual stage filter, control electronics and relay interface.

High quality aspirating smoke detector work by continuously drawing air into the pipe network by a highly efficient aspirator. A sample of this air is then passed through a dual stage filter. The first stage removes dust and dirt from the air sample to enter the laser detection chamber for smoke detection. The second ultrafine stage has the unique feature of providing an additional clean air supply to keep the optical surfaces within the detector clear from contamination and to ensure the stable calibration and long life of the detector.

Aspirating Smoke Detection Systems are capable of providing the earliest possible warning of a potential fire and detects fire at very early stage, before smoke detection system. Hence used in critical areas like telecom rooms, hotels, data centers, server room, hub room etc.

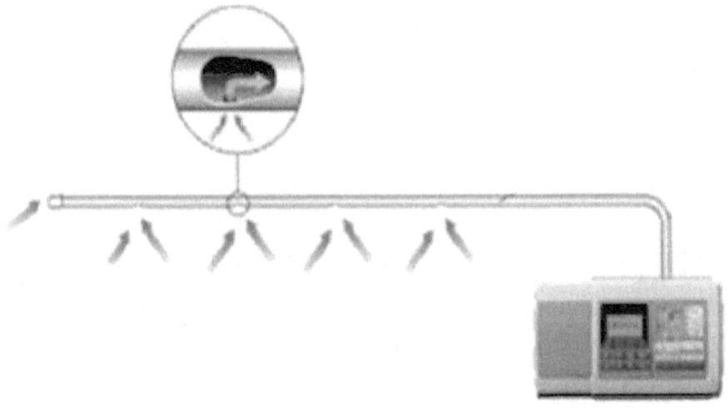

Image by Xtralis

LINEAR HEAT SENSING

1) Linear Heat Detection (LHD) is a line-type form of fixed temperature heat detection used in common commercial and industrial environments. This linear cable can detect a fire anywhere along its entire length and is available in multiple temperatures.

2) These are specialist cables designed to sense heat or sudden increases in temperature anywhere along their entire length.

3) They are designed to be used in commercial and industrial fire alarm systems where conventional heat detectors may be difficult to install, or may not be suitable.

Fire detection in Electrical Panels and Distribution boards

<u>Continuous thermal monitoring</u>- In electrical panels or in distribution boards electrical fires starts due to faulty connections and due to which temperature of busbars, cables, connections etc. starts increasing. Subsequently either fault happens which gets detected and equipment trips or fault remain unnoticed and causes major fire. In view of this deviation in temperature happening from normal operating conditions is required to be detected quite before equipment failure. There are two solutions- one is infrared thermography and second is continuous thermal monitoring.

In case of <u>infrared thermography,</u> thermal camera is used to inspect generally once in a year or as per predetermined frequency. Due to which there is risk of not getting information in time or well before temperature increases and also in some cases to take camera is not feasible. Hence Continuous thermal monitoring becomes more useful. As in case of thermal monitoring temperature is monitored continuously and reporting is done faster along with generation of alarms in case of abnormal temperature rise.

In case of <u>continuous thermal monitoring,</u> sensors are installed permanently installed on individual important connection points. This is best suited for low-density applications with few connection points. These sensors give signal wirelessly on centralized monitoring system. If sensor reading shows an abnormal temperature rise at any connection point, an alarm will be generated at the software level. The software is configured to send a notification to the mobile devices of maintenance teams.

Wireless sensor

There are smart sensors available for early detection of overheating. In case overheating is happening in cables, busbars or wire connections then it analyzes gas and micro particles in the air and sends alerts. As such electrical switchboards can be saved and fire incidences can be stopped. Temperature Measurement range of sensor is 15 °C to +70 °C. It also measures air quality index, temperature and humidity. This works with software to give alerts to users. This is suitable to use in distribution boards, panels.

Photo from Schneider Electric

FIRE EXITS

1) Fire exit is very important topic to be considered at the time of design as per standard and maintained very well. Fire exit is evacuation passage for occupants to

go out of fire zone safely in case of fire. Door should be openable outward.

2) Exit should not be locked at any time and should have panic bar assembly which will get operated to open door in case pressed.

3) Fire exit comprises exit door, exit passage, fire exit staircase which will open at safe place like on street, refuge area, terrace.

4) Doors leading to fire exit will be fire resistant for 120 minutes to be used. It is recommended to have passage height of 2.4 meters and door of 2 meter height.

5) Number of exits will be decided by occupant load factor as mentioned in standard.

6) In case of fire, smoke tends to spread in all direction including stairways. So, stairway will have opening or cutouts, so that smoke can go out easily in case of fire. In case there is no opening for smoke to extract naturally pressurization of stairway is done. This will ensure smoke will not be present in stairway from fire area in case of fire. And occupants can safely go out and need not have to go out through smoke.

7) Fire exit will not be used for storing any materials and should be essentially free.

8) Lift cannot be fire egress passage and in fact in case of fire lift should not be used.

9) Emergency lighting with 90 minutes back up time to be used in passage.

10) EXIT will be clearly marked by self-glow plate or lights.

11) In case of access control doors which are in exit path, will have manual release device. This can open the door manually if need arises. At the same time the doors will be opened automatically as soon as fire detection or sprinkler system gets activated.

STAIRCASE

1) In case of multi-story buildings incombustible internal or external staircases will be provided for access to each floor.

2) If a compartment has a floor area of less than 500 sq. m. then one access staircase can be proposed. For higher areas minimum two staircases will be needed per compartment.

3) In case the area of the compartment exceeds 2,000 sq. m. then additional staircase shall be provided for every additional 1,500 sq. m. or part thereof.

4) Internal staircases shall be so located that at least one of its enclosing walls is an external wall of the building.

5) A door opening shall be provided in this external wall at ground floor level.

6) Every opening from the staircase to floor shall be fitted with a single fireproof door.

7) External staircases without side covering need not have openings at each floor level protected by single fireproof doors.

8) Staircases shall be not less than 750 mm clear width with treads not less than and risers not more than 200 mm.

9) The staircase will have an inclination of more than 60 degrees to the horizontal.

PRESSURIZATION

Pressurization is a method adopted for protected escape routes against ingress of smoke. Air shall be injected into staircases, lobbies or corridors to raise their pressure slightly above the pressure in adjoining parts of the building. As a result of which ingress of smoke into escape routes will be prevented. The mechanism of pressurization shall be automatic linked to fire alarm system. Manual Operation shall also be possible. The pressurization fans shall be installed on the terrace. Pressurization shall be considered for Lift well, Staircase & Lift Lobby as specified in local norms.

COMPARTMENTALIZATION

It is important to limit the spread of a fire in area. Hence compartmentation in the building is carried out as per standard in any building. For business building the same is given as per standard 3000 sq. mtr.

As such physically separation is made with walls and passage between two compartments are made with fire doors. However, when physical separation is not possible like in open offices or in large halls then fire curtains are generally recommended as an effective means of Fire Compartmentation. With which fire does not spread from one compartment to other and get restricted in typical area where fire incidence has occurred.

Fire Curtain barrier assemblies are technologically advanced fire-resistant fabric barriers encased in a compact steel housing. Barriers remain invisibly retracted until activated by an alarm or detector signal, at which time they descend safely to their fire operational position.

The fabric of fire curtain is 120 min fire resistant at 1000-degree C. Fire Curtain system assembly comprises of control panel, top box, guide rail.

Assembly is designed with Gravity fail safe system. So that even with a total power failure the curtain will still close down into a protection mode.

It has weighted bottom bar to ensure correct operation. The emergency pull back switch is provided on both side of the curtain. The control panel has a backup battery system.

Under normal operating conditions the curtain will be held in the retrack position via the motors operating at stall voltage. Upon activation of the fire alarm the control panel will remove the supply voltage and the curtain shall descend under the power of gravity in a controlled manner.

The Fire Curtain reset automatically when the control panel is reset. The power required for driving the system is taken from batteries provided in fire curtain system. The fire curtain shall comply for codes like DIN, BS, EN etc.

A Fire Curtain. Image from www.coopersfire.com

FAÇADE REQUIREMENTS

Building envelop is covered with façade glass and making it closed building. But in case of fire, it becomes dangerous situation because smoke cannot go out and suffocates occupants. In view of this standards have asked to have part openable windows. So that in case of fire the same will opened and will let smoke out. At times glass panels gets detached due to fire and starts falling down. This causes indirect injuries for people at ground level.

Accordingly, as a rule facade glass at each floor shall have means to open at least 10 percent of the total facade area to exhaust smoke during emergencies.

Such openings, with sill level at 1.2 m, shall be in the form of openable panels of size not less than 1500 mm X 1000 mm opening outwards.

For fully sprinklered building having fire separation of 9 m or more, tempered glass in a non-combustible assembly, with ability to hold the glass in place, shall be provided.

It shall be ensured that sprinklers are located within 600 mm of the façade glass providing full coverage to the glass

Automatic opening of windows can help preventing the dangers. Automatic system is composed of electric window opener, controller and smoke sensor. The sensor after detecting fire sends a signal to the controller to drive the motors opening the window automatically. The smoke ventilation window can be opened manually also by pressing the fire alarm button. So that in case of fire instead of waiting to operate the system through controller occupants can operate the window manually also.

Exit SIGNAGE

Exit signages are provided so that occupants can find out way to nearest emergency exit in case of fire or emergency evacuation. Exit signs will have direction of exit. Signages will be well illuminated constantly with alternate source of supply. The color of the exit signs shall be green. Earlier exit signages are with word EXIT only. But now a days it is in pictogram format with or without text. Variants in exit signages are with lithium-ion battery powered illuminated signages with battery charger. So those batteries will be always charged through building supply. Back up time is normally 3 Hrs and having visibility more than 30 Mtrs.

ESCAPE LIGHTING

Escape lighting will be provided throughout the exit path so that occupants can safely come out of building. Power for the same will be provided from independent source other than normal and shall give back up time of 90 minutes. It is recommended to have 10 lux at center line. Lights will come on within 5 seconds after failure of mains power. Lights will be provided in exit passage, near to staircase, near to firefighting equipment, at exit and at intersections.

Escape lighting can be achieved by (a) Central UPS and battery banks (b) localised power packs provided to lights (c) Centralized battery system.

First and second will be used for smaller installations. But in case of large installation CBS central battery system solution is now a days provided.

The CBS central battery system is state-of-the-art, reliable and easy-to-operate system. The system monitors each output circuit individually with the use of address modules for each luminaire. The system allows flexible mode of configuration. Using the driver one can programme the mode of the address module without interfering into the luminaire and without any specialized service software.

CBS comprises of Central control console with control module, battery control modules, charge module, battery etc.

LED Luminaires for escape lighting, exit signages.

Monitoring modules having address code for common switching of the Emergency Lighting luminaires located near lighting distribution board.

Communication cable interlinking monitoring modules with Central control console.

Power cable between Luminaires with Central control console.

Power fails or power supply is cut off then monitor modules gets signal and communicate to Central control console. Central control console interns gives supply to luminaires. And emergency lights are ON. And people can evacuate safely.

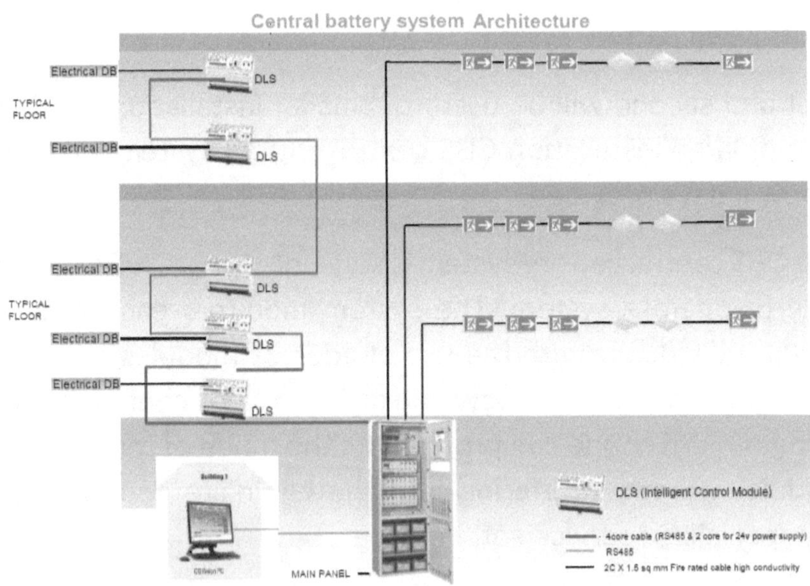

Photos from Eaton

Public address system

Public address system is essential part in case of fire which gives emergency evacuation message in fire condition.

PA system would be consisting of –

- Ceiling mounted speakers which are surface type or recess type. It is generally polypropylene Speakers with line matching transformer having power handling capacity of 5 watts or more as per usage complete with cabinet.

- Amplifiers

- Paging system

- Integration provision to integrate PA system with fire alarm system.

In case of emergency a single channel would be activated which can serve the purpose for repeating emergency messages.

Offices are typical examples of applications with a relatively low number of zones, typically one for each floor in the building plus a few extra for the parking garage, reception, stairwells and so on. Each zone in the office levels has a relatively low output power requirement per zone since working areas are quiet environments.

Offices require an essentially emergency evacuation system for evacuating staff in case of fire and also used for additional usage such as paging personnel and background music system at selective location and addressing employees from one location if required.

Offices are having ceiling mounted concealed type speakers, parking areas can have horn loudspeakers, stairwells can have wall mounted loudspeakers and spaces like atrium can have high ceilings and so line array column loudspeakers.

After seeing life safety we can now see methods of fire suppressions.

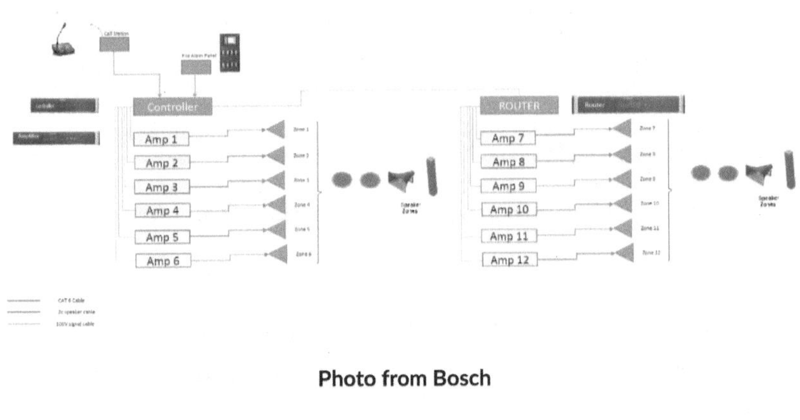

Photo from Bosch

FIRE SUPPRESSION

Fire suppression is carried out with different methods. Very commonly used are fire extinguishers.

Fire Extinguishers Specification and Type

Fire extinguisher shall have general specification as under-

1. Fire Extinguisher should be ISI marked as per IS 15683

2. Cylinder body shall be deep drawn two-piece construction made out of Mild Steel CR-2 with the cylindrical bottom skirt ring.

3. Cylinder shall be certified from approved laboratory for minimum 85 bar burst pressure and the body shall withstand minimum burst pressure of 80 bar at welds.

4. Internal and external surface of cylinder shall be pure polyester powder coated for corrosion resistance.

5. Valve shall be made of forged brass with nickel chromium plating. Pressure release device shall be provided on the valve to protect the extinguisher from over pressure. Valve ring shall have minimum clearance of Min. 2 mm from the Cylinder dome. Valve shall withstand minimum burst pressure of 80 bars for 60 sec.

6. Safety pin of Stainless-Steel grade SS 304 shall be provided with lead seal.

7. The pressure gauge port shall have a check valve to enable removal of pressure gauge without loss of cylinder pressure.

8. The Gas cartridge shall be of the same brand as extinguisher and ISI marked (IS: 4947)

9. Siphon Tube and Strainer shall be made of stainless-steel SS-304 grade with thickness not less than 1.8 mm.

10. The hose shall be of EPDM material with minimum bursting pressure 40 Kgf/Cm2, with minimum length of 450 mm.

11. The nozzle shall have metallic inserts for thread protection and shall have a minimum length of 100 mm.

12. Cylinder body shall be permanently marked UV resistant label with brand, year of manufacturing, test pressure and operating temperature range.

13. The cylinder serial number, manufacturing date & batch number shall be printed on cylinder.

14. Each cylinder should have vinyl sticker with operation instructions.

15. 10% spares quantity to be kept in stock.

16. Extinguishers shall be checked once a week and to confirm that all movable parts such as plunger, nozzles, are working properly. There are no leakage

and maintaining the pressure. Extinguishers are to be removed from site in case of any leakages are observed.

Fire Extinguishers Types

(All images of Fire extinguishers are taken from Kanex catalogue.)

Powder extinguishers

Stored powder pressure dry powder fire extinguishers are high performance used for burning materials, petrol, oil, electrical equipment. It has Mono Ammonium Phosphate based dry chemical powder. It has brass nickel plated head valve, EPDM rubber braided flexible hose, rechargeable, squeeze operation. It is used for Class A, B, C and electrical equipment fire. These should not be used in enclosed spaces as the dry powder can be easily inhaled and also it's not easy to clean up the leftover residue once the fire is over.

Clean Agent extinguishers

Clean agent FE 36 non-corrosive electrically nonconductive free of residue fire extinguishers used for high value equipment. It has high performance extinguishing agent with zero ozone depletion, maximum visibility during discharge and no thermal or static shock. It has pure polyester powder coating and has EPDM rubber braided flexible hose, rechargeable, squeeze operation. It is used for Class A, B, C and electrical equipment fire.

Clean Agent SS Body extinguishers

Clean agent FE 36 non-corrosive electrically nonconductive free of residue fire extinguishers used for high value equipment. It has high performance extinguishing agent with zero ozone depletion, maximum visibility during discharge and no thermal or static shock. It has smooth and polished body and has EPDM rubber braided flexible hose, rechargeable, squeeze operation. It is used for Class A, B, C and electrical equipment fire.

Carbon Dioxide type extinguishers

CO2 is non-conductive fire extinguisher used where there is high electrical risk like transformer, generator, panels etc. It is available in in squeeze grip and wheel type. It has body made up of seamless manganese steel duly enamel painted, brass head valve and rechargeable type. It is used for Class B, C and electrical equipment fire.

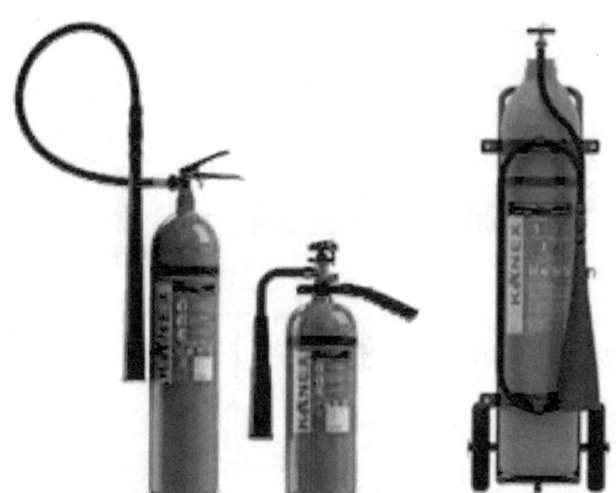

Foam and Water type stored pressure extinguishers

Pressurized foam and water fire extinguishers for fires due to volatile liquids and free burning materials like paper, cloth, wood, furniture etc. But not to be used for electrical fires. Water based suitable for Class A fire and Foam type suitable for Class A and B. It has pure polyester powder coating and has EPDM rubber braided flexible hose, rechargeable, squeeze operation.

These extinguishers work by creating a blanket type cooling effect on the fuel that is responsible for causing the fire. Foam creates barrier between burning liquid and flame. Generally used in premises, warehouses, residential properties, hospitals, schools, offices and buildings storing flammable liquids.

Same are available in stainless steel body also.

Powder extinguishers cartridge type

It is Sodium bi carbonate based dry powder fire extinguishers suitable for freely burning materials, petrol, oil and electrical equipment. It is used for Class B, C and electrical equipment fire. Whereas Mono Ammonium phosphate based dry powder fire extinguishers suitable for Class A, B, C and electrical equipment fire. It has pure polyester powder coating and has EPDM rubber braided flexible hose, rechargeable, squeeze operation.

Kitchen Fire extinguishers

Kitchen fires are classified as K class as per American system and F class as per European and Australian systems. Kitchen fires occurred due to cooking oil, greases and animal fats. It creates foam layer on the surface which holds the vapor and steam and extinguishes fire. It has smooth polished stainless-steel body, EPDN rubber braided discharge hose, discharge nozzle made up of aluminium allows high volume low velocity delivery.

Metal D Class fire extinguishers

Class D fire extinguishers are designed to tackle flammable metal fires. In such fires water and foam are not suitable. It contains blended Sodium Chloride based dry powder extinguisher agent which is well used for all metals except lithium and its alloys for which TEC powder is used.

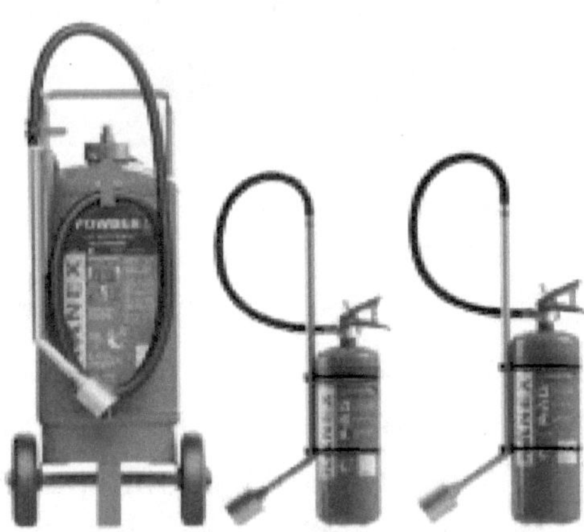

Different type of fire extinguisher techniques for different class of fire is given bellow-

	Fire Classes					
	Class "A"	Class "B"	Class "C"	Class "D"	Electrical	Class "K"
Fire Extinguisher Type	Combustible Material (e.g. Paper & wood)	Flammable Liquids (e.g. Paint & Petrol)	Flammable Gases (e.g. Butane & Methane)	Flammable Metals (e.g. Lithium & Potassium)	Electrical Equipment (e.g. Computers & Generators)	Deep Fat Fryers (e.g. Chip Pans)
Water	✓	✗	✗	✗	✗	✗
Foam	✓	✓	✗	✗	✗	✗
Dry Powder	✓	✓	✓	✓	✓	✗
CO2	✗	✓	✗	✗	✓	✗
Wet Chemical	✓	✗	✗	✗	✗	✓

Other than fire extinguishers other fire suppression methods are which are as given below.

Water based fire suppression

The fire fighting arrangement shall be designed as per the requirement of local guidelines, NBC, TAC, CFO NOC, UL, NFPA, FM as the case may be, to ensure the highest safety standard and uniformity of system. The fire protection shall be fully operated, tested and approved by competent authority under simulated conditions to demonstrate compliance with the most stringent standards, codes and guidelines before property is opened for occupants.

Standard has given three classes for purpose of hydrant design as - Light hazard, Ordinary hazard and High hazard. Exhaustive list of type of factories, manufacturing units, hotels etc. are given in standard. Accordingly, water supply, pumping capacity and other details are required and designed. In general requirements are discussed below-

System comprises of mainly following functional system -

Fire water static storage:

Water Source -

1) Effective capacity of firefighting tank shall be more than 2 hrs. to 3 hrs as instructed. Storage shall be provided at basement in accordance to CFO NOC requirements.

2) The storage of general water supply will be more than 1,00,000 litres.

3) One static tank shall be provided for buildings with light hazard occupancy and two for buildings with ordinary hazard occupancy.

4) The storage tank shall be provided with a 150 mm fire brigade pumping connection to discharge at least 2,275 Litres per minute into the tank.

5) This connection shall be above 150 mm above the overflow level of the tank.

6) The fire brigade connection shall be fitted with four numbers of 63 mm accessible inlets at a suitable position at street level.

7) 150 mm connection shall be taken from the four 63 mm instantaneous inlets direct to each hydrant riser so that the fire brigade may pump to the hydrants in the event of hydrant pumps being out of the commission.

8) Water for the hydrant services shall be stored in an easily accessible surface / underground lined reservoir or above ground tanks of steel concrete or masonry.

9) Reservoirs of and over 2,25,000 litres capacity shall be in two interconnected equal compartments to facilitates cleaning and repairs. The fire tank must be in two compartments with a baffle wall in between two tanks. Wall is not going right down to the base of the tank so that water is not stagnant.

10) Static fire water storage tank with separate set of pumps for Fire Protection System shall be provided in the underground tanks.

11) Also there will be tank on terrace, acting as reserve storage. It will act as water source for the first aid hose reel system.

Hydrant system:

1) It comprises of internal which is inside the building and external means outside of building, hydrant system.

2) Both internal and external fire hydrant system shall be provided with accessories such as landing valves, hose reel, first aid hose reels, complete with instantaneous pattern short stainless steel branch pipe and nozzle.

3) In case of internal hydrant system there will be vertical fire shaft. It will have lockable MS door with glass, painted RED and FIRE written on it on every floor. All the accessories mentioned above shall be housed inside the Fire shaft on every floor. If hose reel, first aid hose reel are placed outside of fire shaft then same will be in MS hose box cabinet with glass painted in RED and FIRE of suitable size shall be provided.

4) In case of external hydrant, same shall be located within 2 m to 15 m away from the building to be protected such that they are accessible and may not be damaged by vehicle movement.

5) At every 30 Mtrs outlet of hydrant will be taken out, where MS hose box cabinet painted in RED and FIRE with glass cover comprising of hose reel will be provided.

Wet Risers

1) Wet risers are a pipe permanently charged with water under pressure.

2) Pipes are throughout the full height of the building.

3) The wet risers shall be located within the lobby near to staircases.

4) The diameter of the riser pipes shall be minimum 150 mm.

5) One or two landing valves shall be connected to the riser pipe at each floor.

6) Operating pressure shall be 7 kg/cm2.

7) The system shall be so designed that shall be kept charged with Water all the time under pressure and capable to discharge 2850 Lts /min at the ground floor level outlet and minimum 900 Lts/min at the top most outlet.

8) The internal hydrant riser shall be tapped from the external main fire line and will be terminated with air release valve at the highest point to release the trapped air in the pipe network.

9) Four-way fire brigade inlet connections at ground level at suitable location will be provided. This will be used by fire brigade to fill the rising mains of hydrant system in case of failure of pumps.

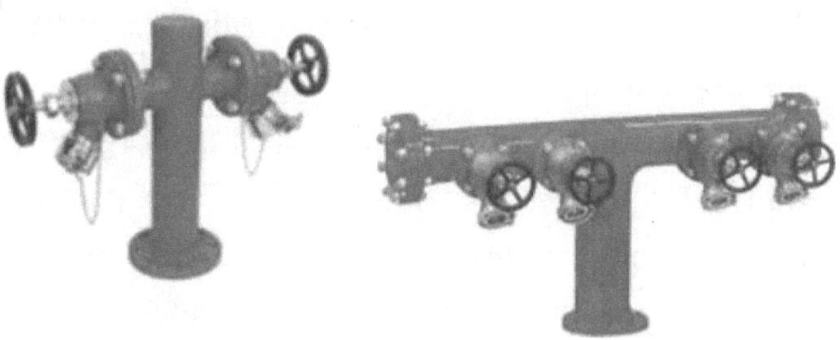

Images by Newage

Hose Pipes and Nozzles

1) 63 mm canvas Hose pipe will be of in two lengths of 15 M each with couplings shall be provided at each hydrant point on risers.

2) Reinforced rubber Branch pipe of 19 mm size 30 mtr long having nozzle fitted shall also be provided.

3) Hose pipe and branch pipe shall be kept in wall box with glass near hydrant.

Images by Newage

Fire Pumping system:

1) Pumping system comprising of independent pumps for hydrant, sprinkler & jockey application (As per Local CFO NOC) to be provided.

2) The fire pump shall be horizontally mounted. It shall have a capacity to deliver and developing adequate head so as to ensure a minimum pressure at the highest and the farthest outlet. The pump shall be capable of giving a discharge of not less than 150 per cent of the rated discharge and at a head, not less than 65 per cent of the rated head. The shut off head shall be within 120 per cent of the Rated head.

3) The pump casing shall be of cast iron and parts like impeller, shaft sleeve, wearing ring etc. shall be of non-corrosive metal like bronze/ brass/ gun metal. The shaft shall be of stainless steel. Provision of mechanical seal shall be made.

4) Bearings of the pump shall be effectively sealed to prevent loss of lubricant or entry of dust or water.

5) The pump shall be provided with a plate indicating the suction lift, delivery head, discharge, speed and number of stages.

6) The pump casing shall be designed to withstand 1.5 times the working pressure.

7) Auto-start pumping set will supply water to each wet hydrant.

8) A stand-by pump set of identical pumping capacity shall also be provided.

9) Pumps shall have capacities of 38 Litres per seconds (137 m3 per hour) or 47 Litres per second (171 m3 per hour).

10) Jockey pump shall be installed in addition to the main pump set. Jockey pumps shall compensate, pressure drop and line leakage in the hydrant and sprinkler installation. It will start at 0.35 Kg/cm2 below normal pressure and stops at normal pressure.

11) The main fire pump shall start at 1 Kg/cm2 below the normal system.

12) The power supply to the fire pump shall be independent of all other supplies within the premises. Means when the power supply to the entire premises is switched off, the supply to the fire pump and other essential equipment shall remain uninterrupted.

13) Pump room preferred to be located at 6 mtr away from building in the compound of the. In case same is not possible then it can be placed in building. Where it is not feasible the pump room can be located inside the building but will have access from outside of building direct in pump room.

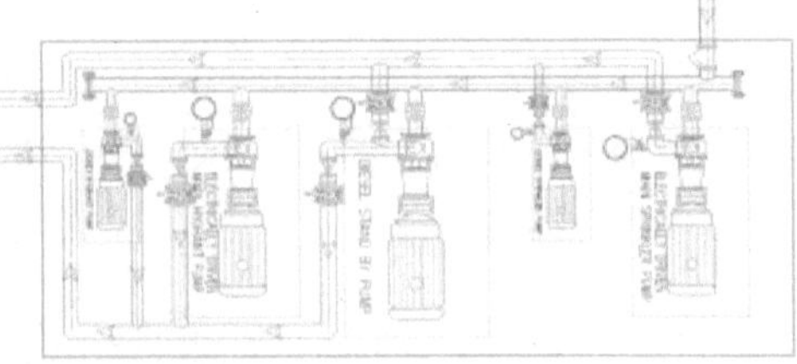

Images by Author

14) Pumps shall be exclusively used for fire fighting purposes.

15) Pumps shall be direct-coupled and not belt driven.

16) The pump delivery pressure will need to be 7 kg/cm

17) Pumps shall not be installed in open. The pump rooms shall normally have brick/concrete walls and non-combustible roof, with adequate lighting, ventilation and drainage arrangements.

18) DG backup for fire pump shall be provided for ensuring operation & performance of the system in case of total electrical power failure.

19) Delivery lines from various pumps shall also be connected to a common header in order to ensure that maximum standby capacity is available. The sprinkler pump shall be isolated from the main discharge header by a non-return valve so that the hydrant pump can also act as standby for the sprinkler system. The ring main shall remain pressurized at all times. To make the system fully functional required automation shall be provided.

20) When the Sprinkler or hydrant valve or hose reel opens, the pressure drops in respective piping network. This will be sensed by pressure switches installed on delivery header of pumping system and gives command to make ON starters of the jockey pumps automatically. With this water requirement is met. In case water requirement is more and pressure drops further, the main pump will start automatically to meet the water requirement.

21) Arrangement is made such that Jockey pump starts and shut down automatically as per the pressure gauge settings.

22) Whereas main fire hydrant, sprinkler or diesel engine driven pump starts automatically but shut down has to be done manually.

23) In case of failure of electrically operated pump then diesel engine driven pump starts automatically.

Typical arrangement is as below-

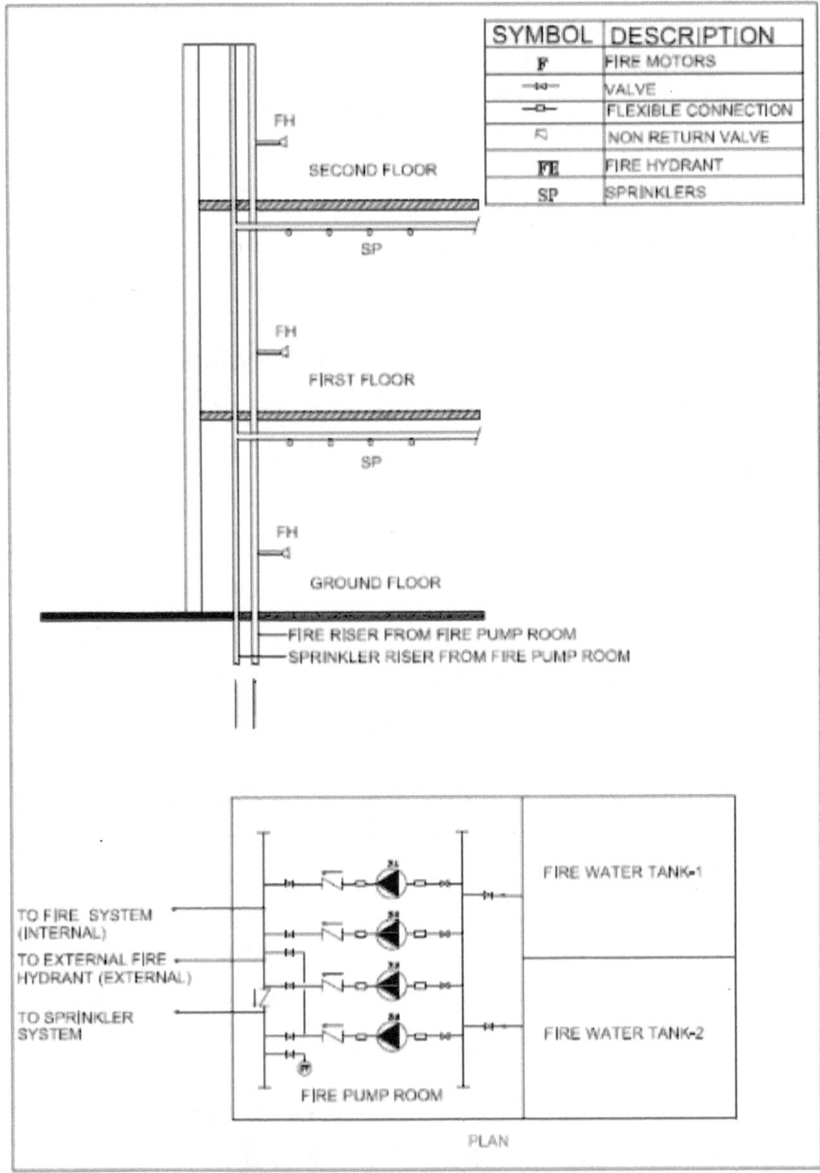

Sprinkler system:

1) In case of high rise building where sprinklers are required sprinkler pumps are provided at ground level in pump room. From outlet of sprinkler pump, sprinkler header is taken out at each floor. From Sprinkler header branch piping is done throughout the floor and sprinklers are provided to it.

Images by Tyco

2) For sprinkler system pipes shall galvanized mild steel welded up to 150 mm diameter will be used. The size of pipe will vary from 25 mm to 150 mm to suit the hydraulics of the system.

3) Quartzoid Bulb Automatic Sprinkler will be used in sprinkler system. Sprinkler heads shall be made of Gun metal and quartzoid bulb sufficiently strong in compression to withstand any pressure, surge or hammer likely to occur in the system. The yoke & body shall be made of high-quality gun metal. The deflector of suitable design shall be fitted to give even distribution of water over the area commanded by the sprinkler.

4) Sprinklers shall be rated for 13 kg/cm2 and factory tested for 3.5 kg/cm2

5) The bulb shall contain the liquid having a freezing point below any natural climatic figure and a high coefficient of expansion. The Operating temperature of bulb shall be 68 degree C.

6) This system is an active fire protection method widely used worldwide. It has further divisions such as –

<u>Wet pipe-</u> In these pipes are charged with water. In case of fire sprinkler bulb breaks open and water comes out through sprinklers.

<u>Dry pipe-</u> In these pipes are not charged with water but are pressurized. In case of fire sprinkler bulb breaks open and air pressure in the piping drops. Due to differential in pressure water enters the piping system and from opened sprinkler it comes out.

<u>Deluge valve system-</u> It is type of dry pipe system except sprinkler bulbs without temperature sensor. Deluge valve is holding the water to come in piping system. As soon as external fire smoke detection system gives signal, deluge valve operates and water comes out of sprinkler head for suppression.

<u>Pre action System-</u> Pre action system is mixture of wet and deluge system. In which there is one variant of single knock system. In which once smoke detection system senses fire, pre action valve opens and water enters piping. And after sprinkler bulb breaks due to increase in temperature water comes out

of sprinkler system. Second variant is double knock system. In which when signal from fire detection system goes to system and pressure drop happens due to opening of sprinkler bulb, then only pre action valve opens.

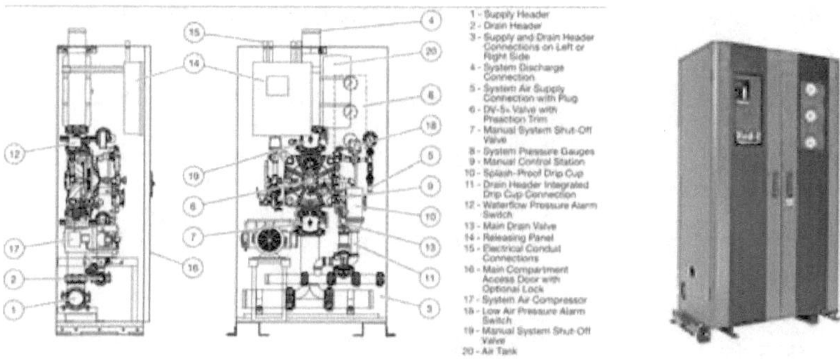

Images by Tyco

Gas based fire suppression

In this fire suppression happens through gas which are either clean agent or inert gases or chemical agents which extinguishes fire. The system consists of gas cylinders, fire detection system, piping, valves and nozzles. In case of inert gas there is absent of chemical reactions such as oxidation and hence there is no damage to property. Few gases or chemicals are- Argon, Nitrogen, FM200, NOVEC 1230, Carbon dioxide, ABC Dry chemicals, potassium carbonate etc.

Chemical Clean Agents – Include HFC-125, FM-200, Halons, Novec 1230 etc. They are fast-acting, human safety and no damage to assets like computers, servers etc. However, after Kyoto protocol FM, HFC are almost unused by all. So also because of ozone depletion potential Halon was banned.

<u>Inert Gases</u> – Include nitrogen, argon and CO2. These gases are safe for both people and the environment. Except CO2 which is unsafe for human so used either in outdoor or where human intervention is absent.

Out of all NOVEC 1230 is used commonly because of -

- It is colorless, nontoxic and odorless clean agent.

- Stored in cylinders in its fluid form.

- It instantly vaporizes upon discharge, totally flooding area.

- It evaporates quickly without harming any valuable assets.

- Most suited for areas where electrically conductive medium is present.

- Effective on class A, B and C fires.

- Zero ozone depletion.

- Very less global warming potential.

- System pressure is less.

- Since they are in liquid stage hence less space required than other clean agent.

<u>Fire suppression system with Gas Cylinders</u>

System comprises of fire detection, gas cylinders outside of protected area, control panel, nozzles in protected area, actuators, valve, manifold etc.

Image by Author

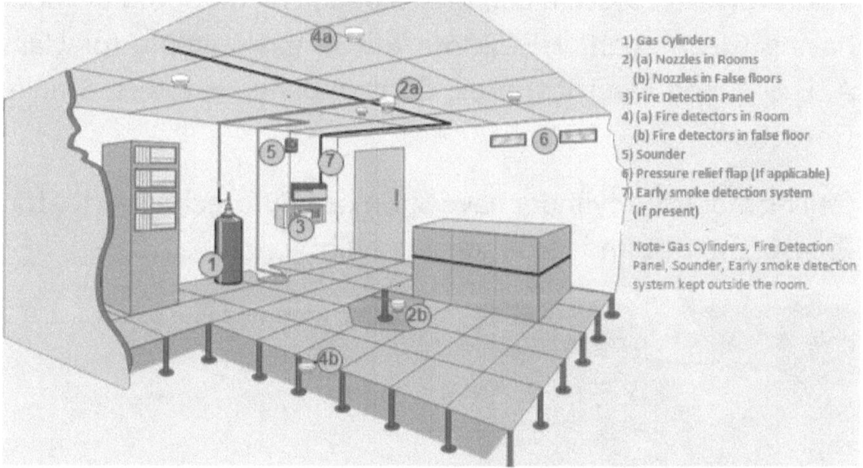

Photo from Minimax

In case of fire, smoke detectors which are in two cross zones, detects the fire and give signal to gas release control panel. During first zone, panel sounder operates and gives alarm to attend fire. In case fire is not extinguished and increased and then second zone also activates, then sounder operates with continuous tone indicating time to evacuate. At the same time panel gives command to gas release module and timer starts. And after predetermined time, gas releases in to protected area. There is facility of manual operation for the entire system also available with manual release switch. Also abort switch for gas not to release is available. Entire system is automatic and is used widely for server room or any other critical room.

Fire suppression system with Ceiling mounted Gas Cylinders

This is variant of above system for small unmanned Room Protection such as computer room, UPS, electrical room, laboratory etc. System comprises of ceiling mounted cylinder having Clean agent HFC236fa/FE-36 gas suitable for class A, B & C and Electrical fires, Control panel, Hooters, Smoke Detectors.

Ceiling mounted cylinder have Sprinkler, Pyrotechnic Actuator, Pressure gauge and Heat sensing bulb.

Ceiling Mounted Fire extinguisher

Image by Author

It can start automatically when detectors detect the fire. It gives command to fire alarm panel, which intern gives signal to actuator heat sensitive bulb breaks open and gas comes out. Same operation can happen in manual mode also. So also. operation happens in self-triggered mode. When fire happens, bulb breaks open and gas is release.

It can detect and extinguish the fire in local area and does not require human interaction or presence. It is available with ABC powder with 2/5/6/9/15 Kg capacities also.

Image from Kanex

Standalone fire suppression system can be used where human traffic is less, in utility rooms, store rooms and in warehouse. They are ceiling mounted and normal fire extinguishers cannot be used. In such cases this will be effective. Ceiling mounted extinguishers can detect rising temperature and beyond the limit it operates. And fire extinguishing media such as ABC or clean agent comes out and extinguishes the fire.

Photo from Ceasefire

Gas based fire suppression may not be feasible in some cases where buildings are in existence and structural changes for installation of cylinders, ducts, pipes etc. or space for keeping cylinders are not possible such as at heritage sites or iconic buildings. So also, day to day activity is stopped and which affects businesses then pre engineering retrofittable solution system can be used.

It comprises of ceiling mounted containers containing the extinguishing agent having nozzles for extinguishing agent, control panel, detection devices. Detection devices detect the fire and send signal to the control panel. Control Panel thereafter sends the signal to the of ceiling mounted containers units for activation. With which the extinguishing agent is dumped into the fire zone and the area is totally flooded with agent which extinguishes fire. Fire extinguishing media can be ABC or clean agent as required.

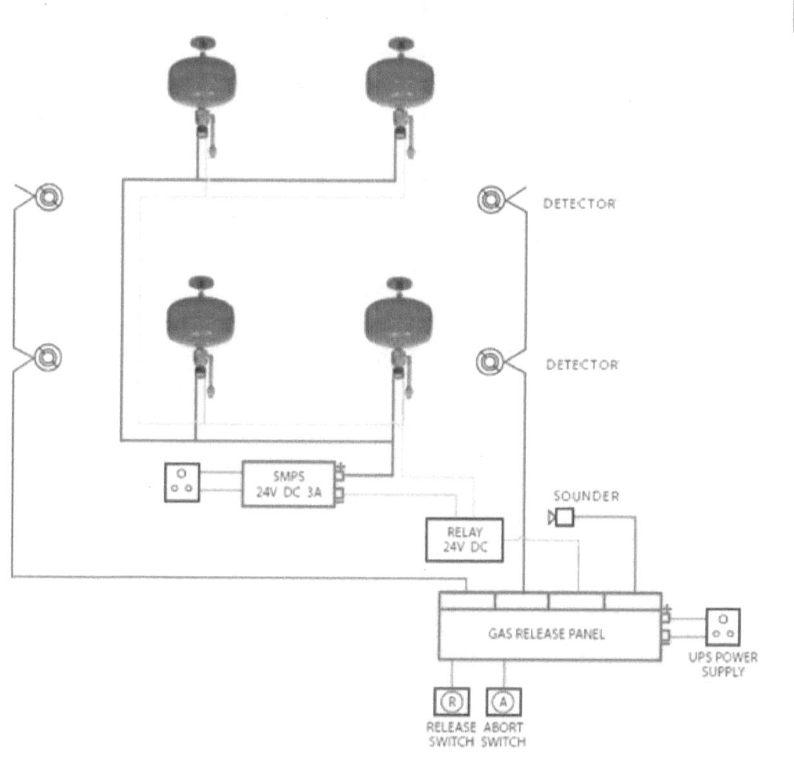

Photo from Ceasefire

Gas based Fire suppression for electrical Panels

It has flexible detection and delivery system called Tubing made out of specially processed polymer materials. It detects heat and also having delivery features. The Tubing, which is pressurized, is placed within an enclosed area of panel wherever fire hazards could be present. Once heat is detected, tube burst and suppression agent discharges from the burst hole. Since it is nearest point of fire it extinguishes fast. Extinguishing mediums can be matched to the particular application. For electrical Novec 1230 is used.

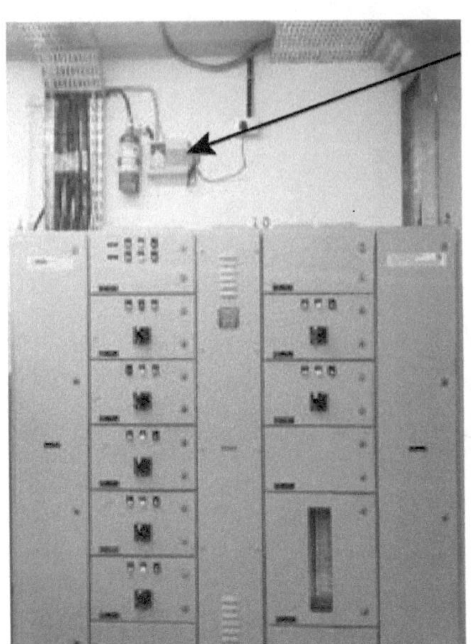

Fire extinguisher for Electrical Panels.

Image by Author

Electrical fires which start from switchgear closets, panels, distribution boards, servers, engine sets, automobiles etc. needs to be controlled immediately else it propagates to large size fire. If at all it is detected still immediate action cannot happen since location is not accessible. By the time action is initiated already damage is done. In such cases Heat- sensitive pneumatic polymer tubing with localized extinguishing agent container can be useful. In this system tube runs throughout risk area such as busbars, cable containment etc. When fire starts and heat-sensitive tube reaches temperature of 150 - 180°C, the tube bursts open at that exact spot and forms a miniature nozzle. Then pneumatic mechanism triggers the valve of the extinguisher and the extinguishing agent comes out of the tube onto the flame and extinguishes fire instantaneously.

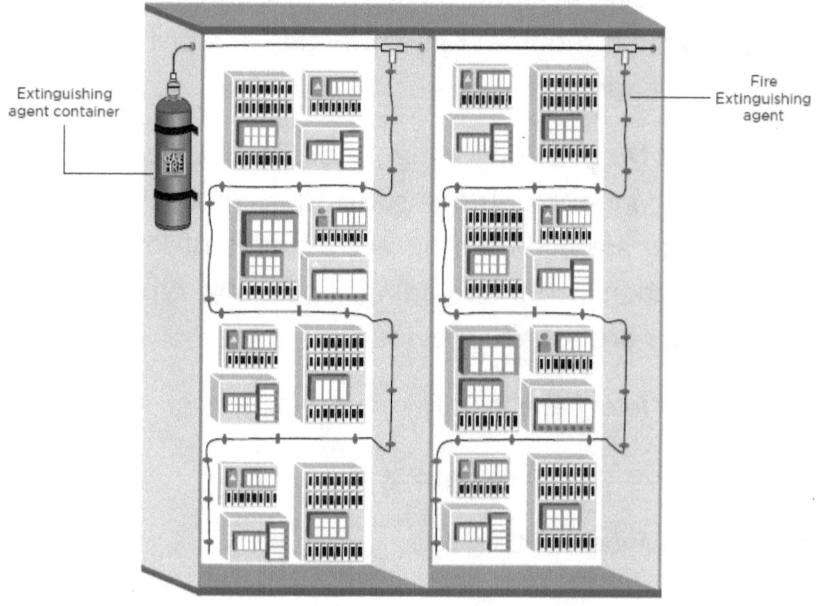

Photo from Ceasefire

Fire Buckets in substation

Fire buckets with round bottom filled with sand mounted on dedicated stand is very useful for small fires and mainly where oil is present. It is used as 'First-Aid' of fire for attacking small fires in their incipient stages and not for large fires. Indicator board for buckets with number and location to be given.

This is provided in addition to foam extinguisher. As per standard same is required to be provided at substation, petrol pump etc. locations.

Portable and Mobile Fire Extinguisher are provided at suitable locations for indoor/outdoor substation applications. These extinguishers are used during early stages of localized fires to prevent them from spreading. Following types of these extinguishers are usually provided.

Pressurized Water Type in 9.01 kg size

Carbon Dioxide Type in 4.5 kg size

Dry chemical Type in 5.0 kg size

Mechanical foam Type in 50 ltrs, 90 ltrs.

Water mist system

A water mist system is a fire protection system which uses very fine water sprays (i.e. water mist). In this either of dry, wet, deluge system can be used. Water comes out as small droplets due to compressed gas pumped through sprinkler pipes. Water mist is water droplets less than 1000 microns. Due to mist the large volume is covered and more heat is absorbed turning droplets in vapour and room get cooled.

The small water droplets allow the water mist to control, suppress or extinguish fires by:

- cooling both the flame and surrounding gases by evaporation

- displacing oxygen by evaporation

- attenuating radiant heat by the small droplets themselves

The effectiveness of a water mist system in fire suppression depends on its spray characteristics, which include the droplet size distribution, flux density and spray dynamics, with respect to the fire scenario, such as the shielding of the fuel, fire size and ventilation conditions.

The low-pressure water mist system shall be designed to function automatically upon activation by an external detection system.

System is fully automatic or manual total flooding fire extinguishing system comprises of - Pump units, Pump control panel, approved deluge valves, butterfly valves, Discharge Nozzles, Distribution pipework.

The low-pressure water mist system is suitable to protect Class B flammable liquid hazards up to 1,280 m^3 and 8 m high in machinery spaces, insulated and non-insulated combustion turbine enclosures, and incidental flammable liquid storage. The machinery space application includes rooms with machinery such as oil pumps, oil tanks, fuel filters, generators, transformer vaults,   gear boxes, drive shafts, lubrication skids, diesel engine drive generators other similar machinery etc.

Few clients are of the opinion that water to be used in fire suppression and not anything else. But due to water there is damage to property especially where electronics is present.

So also, in case where chemicals, metals, foundry or electricity is present it reacts and could lead to permanent damage or explosion or electrocution. Hence few clients are of opinion to use gas or chemical-based suppression system.

The other system like hybrid extinguishing fire suppression system combines water and nitrogen (inert gas) extinguishing agents from a single source. This hybrid system uses a proprietary supersonic technology to atomize the water which allows the nitrogen to combine into a single dense extinguishing material. This special mixture accomplishes two crucial functions: cooling and oxygen reduction and suppresses the fire.

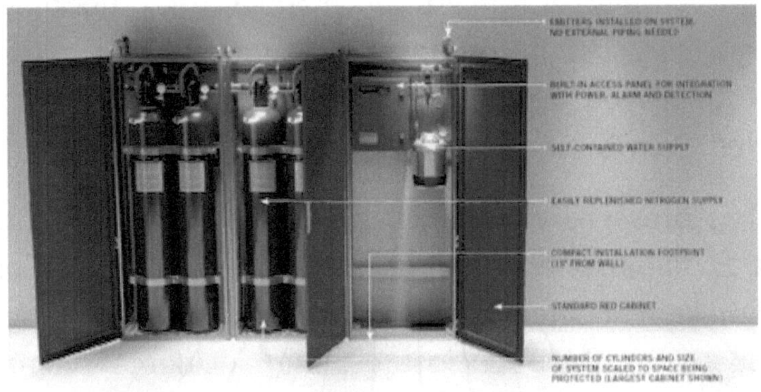

A Hybrid Fire Extinguishing System. Image from www.victaulic.com/

Water Spray System for Transformer

Water spray systems are often used to provide fire protection for oil-filled electrical transformers. High velocity water spray system is one of main water-based fire protection system which is installed to extinguish fires involving liquids with flash point 65°C (150°F) or higher. Some of the areas where this system is widely used are: -   Transformers, Oil Filled Equipment of Power Stations, Turbo-alternators and other Oil-Fired Boiler Rooms, Oil Quenching Tanks.

High Velocity Water Spray System. Image of RGSE Project, Gurugram (India).

Nitrogen Injection System for Transformer

Nitrogen injection systems includes - Nitrogen tank installed outside and away from the transformer. It is connected to the transformer by piping and valves. System also comprises of Fire detectors and Signal Box.

In case of internal fault of transformer temperature of tank increases. Detectors detect the high temperature rise and also Buchholz relay gives the signal. Due to both signals of detectors and Buchholz, Extinction mode gets activated. During this mode Oil from the tank top drains a quantity of oil and also Nitrogen starts injecting in tank. Due drain of oil, pressure in tank reduces. So also, injected high-pressure nitrogen gas to the lower side of the tank, stirs through oil and goes up. Nitrogen gas reduces the oil temperature at the top of the tank and extinguishes the fire.

At the same time excess pressure of the oil / gas is getting reduced and stops tank from rupture.

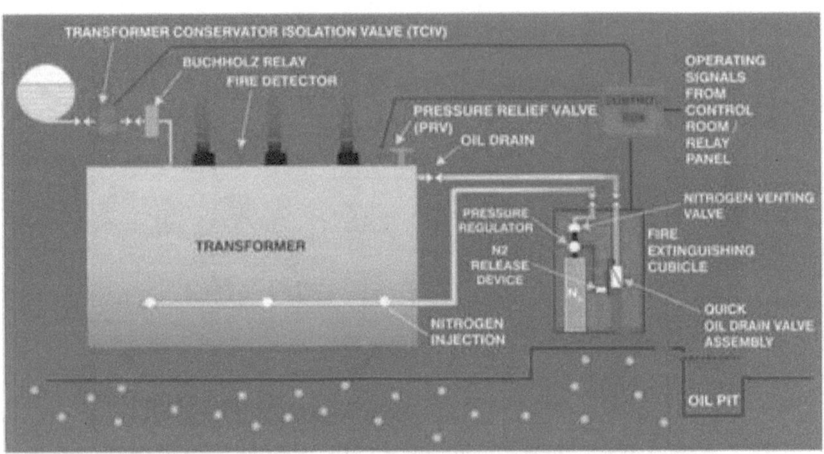

Image from CTR Manufacturing Industries Private Ltd.

CONCLUSIVE REMARKS

"If four things are followed - having a great aim, acquiring knowledge, hard work, and perseverance - then anything can be achieved". - A. P. J. Abdul Kalam

<u>We have to set aim of eradication of electrical fires completely.</u> Our aim should be towards zero fires due to electricity. We have adequate text books of Rules, Regulations and Codes. Now it is our turn to carry out hard and persistent work as per standard without any compromise to achieve this goal.

As per Rules lot of responsibility is assigned to licensed contractor who is directly involved in electrical work. As such prime responsibility lies on him to use standard materials, to carry out work as per codes and there after performing thorough testing commissioning to make installation safe.

Electrical Consultants are involved mainly in large size projects. They play an important role in designing the system and specifying the standard materials, method of good workmanship, quality checks, witnessing the tests to have overall check. Their services and advice should be taken consolidated so that their independent and expert opinion to client will result in healthy installation.

As seen in earlier topics, electrical fires happen due to many reasons and not only through short circuit. Short circuit can be one of the causes, but there are too many other reasons also.

Reasons Could be -

- Nonstandard material
- Improper planning, design, execution
- Inadequate testing
- Not Done Maintenance
- Budget
- Time Frame
- Ignorance
- Human Errors
- Unskilled Staff
- Inadequate Tools
- Material Availability
- Material Failure
- Undue Expansions
- Business Pressures
- Unintentional Work

Reasons can be any, but it is every one's responsibility to take proper care at each stage which can avoid such untoward incidences. Every likely we may come across a situation to say 'yes' compulsorily, when an obvious reply is 'No'. Then collect your strength to say 'no' and bring it into practice. You may be asked to compromise on an electrical subject but keep your foot down and do not budge to compromise on 'electrical engineering' as well as topic of 'fire'.

In case of fire if root cause is found out then take corrective and preventive actions so that such issue will not arise in future. Also, educate yourself from the incident and use this knowledge in future for safety and prevention of fire.

Knowledge of type of fires, fire extinguishing methods and mainly why fire propagates is very important for the people in electrical stream and they must learn about it.

Fire and electricity individually are vast subjects. 'Fires due to electricity' is the most important, relevant and connecting subtopic between the two. If both streams of fire and electricity are bridged together and if fire people understand the electrical part & electrical people understand the fire part, then gap between two will be bridged and resultant will give good results.

Let us work hard on subjects and implement our knowledge in practice so that we remain in safe environment. Working together in unity with support of strong basics and foothold of values and principles will only get desired results. So, it is responsibility of each and every individual to stand by this subject and exterminate electrical fire.

STATISTICS

Indian Fire statistics

National Crime Records Bureau (NCRB) working under 'Ministry of Home Affairs' of Government of India gives data of fire incidences and Loss of lives occurred due to it. As per the data provided Number of cases, Loss of lives, Fire happened in various cases from Year 2015 to 2019 are given below-

	CASES	NO OF CASES					LOSS OF LIVES				
		2015	2016	2017	2018	2019	2015	2016	2017	2018	2019
1	Fire in School Buildings	10	21	6	7	23	9	19	3	7	28
2	Fire in Commercial Buildings	716	463	446	367	328	816	459	382	284	330
3	Fire in Residential/Dwelling Buildings	7493	8359	7660	7241	6364	7445	8478	7614	7208	6329
4	Fire in Government Buildings	35	48	13	24	60	28	41	10	18	54
5	Fire in the Mines	25	28	26	13	2	25	28	26	14	2
6	Fire in Factory Manufacturing Combustible Materials including Cracker/Match Box Factories	410	253	76	51	27	430	306	72	71	33
7	Fire in the Factories (other than Sl.– 6)	212	151	235	184	181	218	168	216	177	177
8	Accidental Fire in Train(s)	58	8	8	9	8	59	6	8	9	7
9.1	Passenger Vehicle (e.g. Bus/Taxi/Auto/etc.)	178	127	85	175	63	187	105	107	155	39
9.2	Private Vehicle	287	163	132	146	136	266	109	108	108	66
9.3	Goods Carriers	254	168	69	181	58	262	155	62	175	45
10	Others	8772	6906	4641	4701	3787	7955	7026	4551	4522	3803
	TOTAL	18450	16695	13397	13099	11037	17700	16900	13159	12748	10915

For better understanding graphical representation can be shown as under-

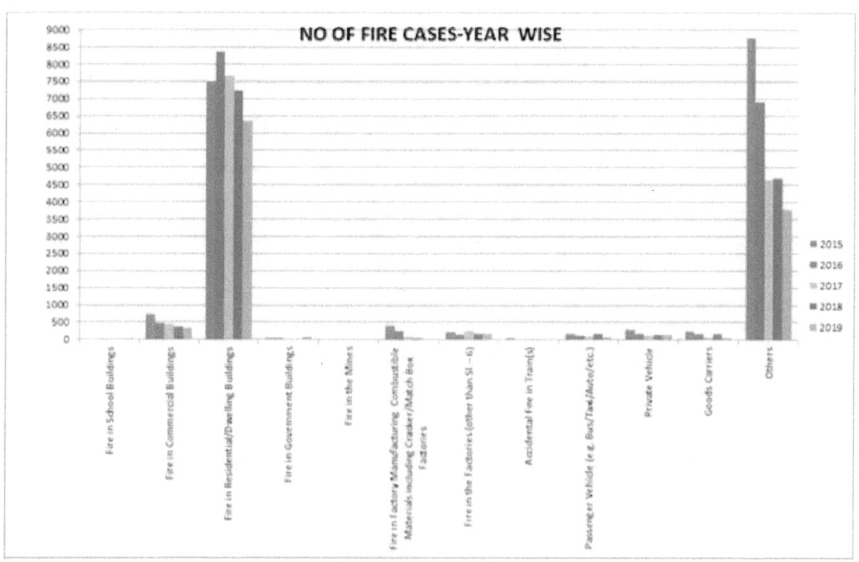

NO OF FIRE CASES-YEAR WISE
Fire in School Buildings
Fire in Commercial Buildings
Fire in Residential/Dwelling Buildings
Fire in Government Buildings
Fire in the Mines
Fire in Factory Manufacturing Combustible Materials including Cracker/Match Box Factories
Fire in the Factories (other than Sl – 6)
Accidental Fire in Train(s)
Passenger Vehicle (e.g. Bus/Taxi/Auto/etc.)
Private Vehicle
Goods Carriers
Others
2015
2016
2017
2018
2019

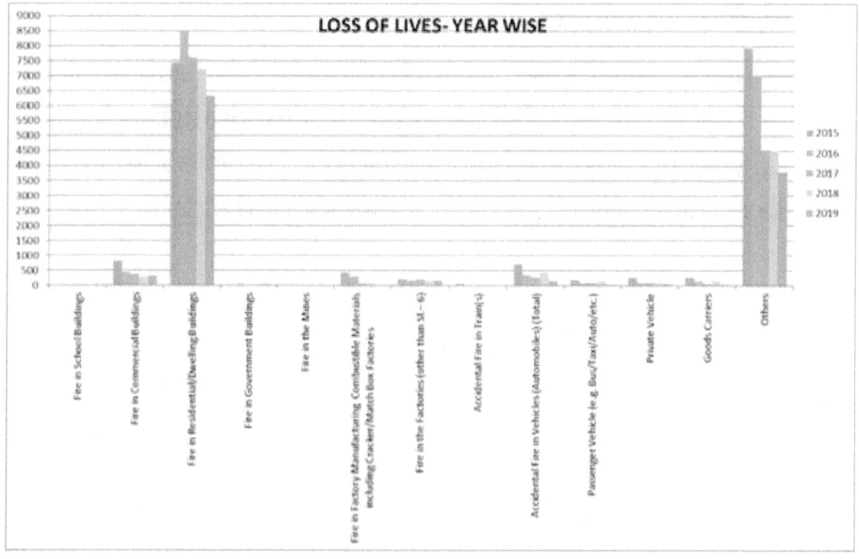

LOSS OF LIVES- YEAR WISE
Fire in School Buildings
Fire in Commercial Buildings
Fire in Residential/Dwelling Buildings
Fire in Government Buildings
Fire in the Mines
Fire in Factory Manufacturing Combustible Materials including Cracker/Match Box Factories
Fire in the Factories (other than Sl – 6)
Accidental Fire in Train(s)
Accidental Fire in Vehicles (Automobiles) (Total)
Passenger Vehicle (e.g. Bus/Taxi/Auto/etc.)
Private Vehicle
Goods Carriers
Others
2015
2016
2017
2018
2019

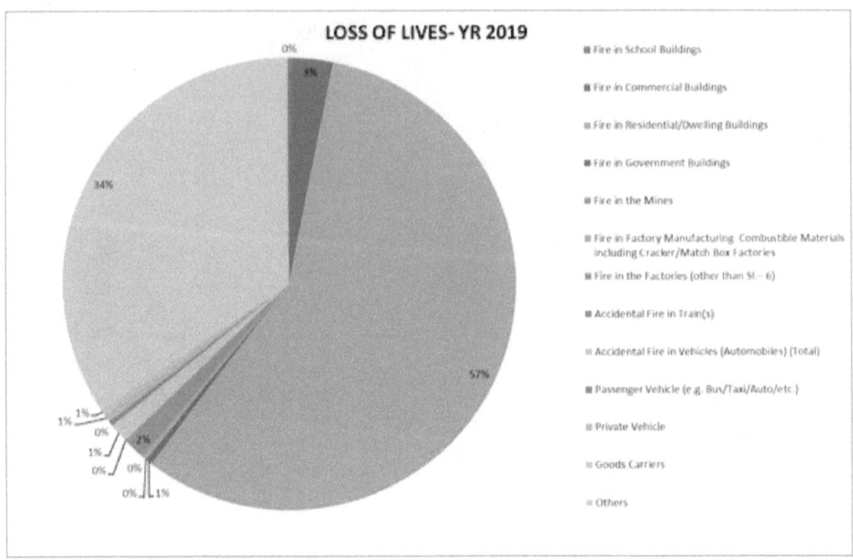

World Fire Statistics

World Fire statistic given by International Association of Fire and Rescue Services CTIF gave World Fire Statistics in 2018 figures, which are given below. Compilation is done on basis of data received by them. Still, it gives worldwide fire cases.

Trends in Fire in the Countries of the World in 2012-2016

	Country	Number of Fires					Average per Year
		2012	2013	2014	2015	2016	
1	USA	1375000	1240000	1298000	1345500	1342000	1320100
2	BANGALADESH	17504	17912	17830	17488		17684
3	RUSSIA	162900	152969	150437	145900	139500	150341
4	JAPAN	44101	48096	43741	39111		43762
5	VIETRAM	1900	2540	2375	2451	3006	2454
6	GERMANY			175354	192078		183716
7	FRANCE	306871	281906	270900	300667	285700	289209
8	GREAT BRITAN	272800	192700	212500	191647	201009	214131
9	ITALY	241232	196196	189375	234675	245727	221441
10	MYANMAR	1219	1673	1629			1507
11	SPAIN	142500	135000	128000	137000	122828	133066
12	UKRAINE	71443	61144	68879	79640	74221	71065
13	POLAND	183888	125425	145237	18447	126228	119845
14	CANADA	45005	37194	36445			39548
15	MALYSIYA	29874	33640	54540	40865	49875	41759
16	PERU	11329	11264	9430	9473	12648	10829
17	NEPAL		1021	958			990
18	TAIWAN	1574	1451	1417	1704	1856	1600
19	ROMANIA	38077			26247	27804	30709
20	KAZAGHSTAN	16145	13621	14477	14452	13952	14529
21	NETHERLANDS			91160	125200	79560	98640
22	GREECE	33731	28232				30982
23	BELGIUM	21369	21228				21299
24	CZECH REPUBLIC	20492	16563	17388	20232	16253	18186
25	SWEEDEN	22657	25392		22785		23611
26	HUNGARY	37106	20177	19536	21056	17534	23082
27	JORDAN	23961	25644	20795	32488	28693	26316
28	BELARUS	34505	7151	7489	7339	5999	12497
29	AUSTRIA	42213	40365	43336	45349	47559	43764
30	SWITZ ERLAND	1434	12893	11658	12477	11803	10053
31	ISREAL	50654	52024			47000	49893
32	BULGERIA	44939	32903	23199	30009	37362	33682
33	SERBIA		22048	16805			19427
34	SINGAPORE	4485	4136	4724	4604	4114	4413
35	DENMARK	14844	15201	14014	12959	13142	14032
36	KYRGYSTAN	3530	4288	4361	4029	3813	4004
37	FIRLAND	11803	13421	14027	11220	12063	12507
38	SLOVASKA	14413					14413
39	NORWAY	7369	7318	8672			7786
40	NEW ZEALAND			10245	10515	10314	10358
41	CROASIA	10857	8062	7317	12156	11143	9907
42	KUWAIT	5609	4661	4889	4914	4771	4969
43	MALDAVA	1984	2033	1890	1816		1931
44	MONGOLIA	3501	3819	4222	4561	3710	3963
45	ARMENIA				6617	4615	5616
46	LITHUNIA	11257	11333	13324	13512	10041	11893
47	SLOVERIA	5570	4175	5917	6983		5661
48	QUTAR	1188	1158	1135	1179	1444	1221
49	LATVIA	8536	9821	12873	11004	9929	10433
50	ESTONIA	4973	5745	6871	5564	5066	5644
51	CYPRUS	6799	8605				7702
52	BHUTAN	45	42	35			41
53	SURINAME			3274	3465		3370
54	BRUNEI	1070	791	1002	1503	1852	1244
55	MALTA				1938	1749	1844
56	ANDORRA	235	212	257	246	274	245
57	LIOCHTONSTEIN	32	46	24	42		36
		3414523	2963239	3191963	3233107	3036157	

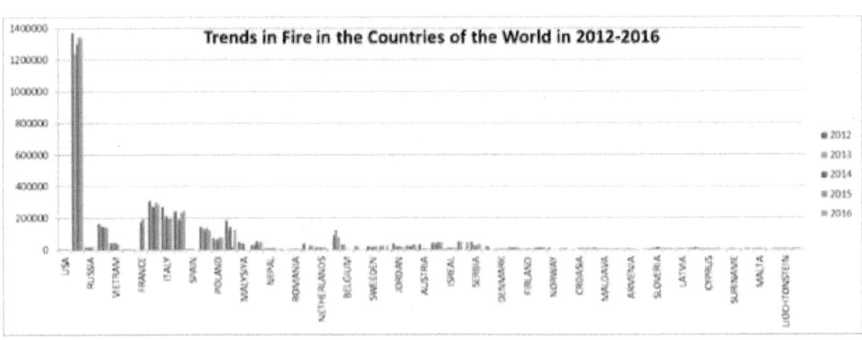

Trends in Loss of Lives in Fire in the Countries of the World in 2012-2016

	Country	Number of Fire Deaths				
		2012	2013	2014	2015	2016
1	INDIA	23281	22177	19513	17700	
2	USA	2855	3420	3275	3280	3390
3	BANGALADESH	210	161	70	68	
4	RUSSIA	11652	10601	10138	9405	8749
5	JAPAN	1721	1625	1678	1563	
6	VIETRAM	78	45	90	62	98
7	GERMANY	384	439	372	367	
8	THAILAND	20	110			
9	FRANCE	362	321	280	335	289
10	GREAT BRITAN	380	350	322	325	
11	ITALY	258	196	141	222	295
12	MYANMAR	184	83	60		
13	SPAIN	170	132	162	143	175
14	UKRAINE	2751	2494	2246	1948	1872
15	POLAND	564	515	493	512	
16	CANADA	149	141	150		
17	MALYSIYA	98	72	139	158	142
18	NEPAL	77	59	67		
19	TAIWAN	142	92	124	117	169
20	ROMANIA	222			646	258
21	KAZAGHSTAN	518	455	401	386	371
22	NETHERLANDS			75	81	42
23	GREECE	49	33			
24	BELGIUM	70	48			
25	CZECH REPUBLIC	125	111	114	115	124
26	SWEEDEN	103	96		110	
27	HUNGARY	140	112	94	108	114
28	JORDAN	42	35	35	52	28
29	BELARUS	927	783	737	578	538
30	AUSTRIA	30	20			
31	ISREAL					19
32	SERBIA		62	73		
33	BULGERIA	53	106	103	109	
34	SINGAPORE	1	4			1
35	DENMARK	65	70	84	68	52
36	KYRGYSTAN	90	80	80	48	80
37	FIRLAND	77	58	86	74	82
38	SLOVASKA	44				
39	NORWAY	40	62	54		
40	NEW ZEALAND				13	19
41	CROASIA	36		21	24	22
42	MOLDOVA	150	120	118	107	
43	KUWAIT	21	17	19	38	50
44	MONGOLIA	75	53	61	59	60
45	ARMENIA					32
46	LITHUNIA	150	160	125	125	101
47	SLOVERIA	8	0	0	3	
48	QUTAR	22	4	18	18	1
49	LATVIA	99	104	94	88	95
50	ESTONIA	54	47	54	50	39
51	CYPRUS	2	5			
52	BRUNEI	1	0	7	4	3
53	LIOCHTONSTEIN	0	0	0	0	
		48550	45678	41773	39109	17310

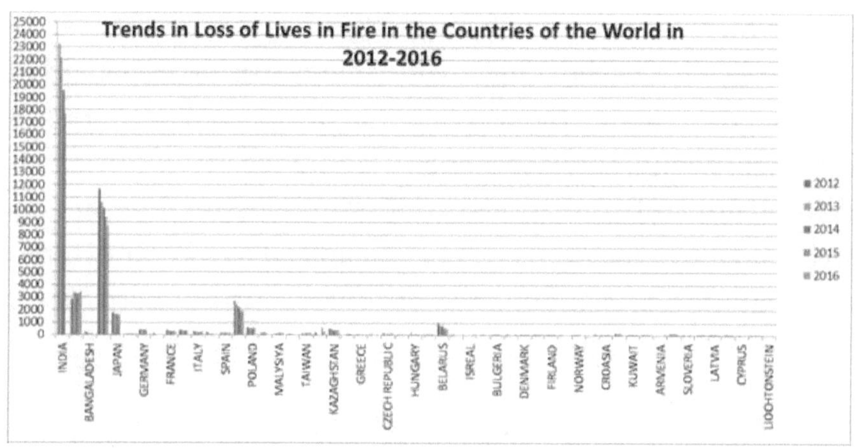

Electrical Fires

Following data is being used to find out how many fires are due to Electricity which may be said as 'Electrical Fires'.

Mumbai city has seen more than 15,000 fires in the past three years, of which about 80% or 12,000 have been caused by short circuits, according to the Mumbai fire brigade. As per newspaper Hindustan Times.

Around 70% of the fire accidents in Allahabad caused by electrical short circuits in both commercial and residential buildings. As per newspaper Times of India.

Over 40% of fire accidents in buildings are caused due to electrical issues- As per newspaper The Hindu.

Nearly 70 to 80 per cent of the fires in the city happened due to faulty electrical connections, as found by the Brihanmumbai Municipal Corporation (BMC). As per newspaper The Indian Express.

About 60% fires are of electric origin on account of electric short circuit, overheating, overloading, use or nonstandard appliances, illegal tapping of electrical wires, improper electrical wiring, carelessness and ignorance etc. It can lead to serious fire and fatal accidents, if proper instructions are not followed. Such incidents can be minimized to a great extent if adequate fire precautions are observed. -As per department of Delhi Fire Services.

Fires Due to Short circuit appeared in Times of India are given in attached annexure list of fires.

50% of fires can be assessed in wiring faults and 25% due to faulty AC. Fires are also seen majority places in electrical meter room, during installation and due to maintenance.

National Fire statistics implies –

Number of fires in 2015 were 18450 whereas in 2019 were 11037.

Figures indicates there is reduction of fire incidences to 40% from 2015 to 2019.

Total Number of Fires in 2019 were 11037 Nos.

Out of which Fires in residential buildings in 2019 were 6364 Nos.

It means Fires in residential buildings are nearly 57% out of total fires.

Balance Nos. are in other type of buildings.

So also, other type of fires are 3787 Nos. which is 34%.

Number of Loss of lives were in 2015 were 17700 whereas in 2019 were 10915.

Figures indicates there is reduction of loss of lives to 61%.

And due to fire nearly 30 deaths per day are happening.

Trend of fires and loss of lives are downwards but still figures are on higher side considering overall increase in development.

Worldwide fire statistics indicates –

There are nearly 3 million fire cases and around loss of lives are 40,000. In comparison to world loss of lives are more in India.

Electrical Fire statistics suggest -

It can be concluded that 70% of fires due to electricity. As such 'Electrical Fires' are 70% of fires taking place.

DIFFERENT FIRE DEFINITIONS

Before going into more details about theory, it will be better to understand what are definitions being mentioned about fire in various references and dictionaries about fire. This will help to understand the fire from the point of view of cause and effects more closely. Because Fire is such a vast subject and unless we go in details at root to differentiate between different types of fires, it will be inappropriate to touch to take subject of 'Electrical Fire'.

1) Fire is the rapid oxidation of a material in the exothermic chemical process of combustion, releasing heat, light and various reaction products. Slower oxidative processes are not included in this definition. At a certain point in the combustion reaction, called the ignition point, flames are produced. The *flame* is the visible portion of the fire. Flames consist primarily of carbon dioxide, water vapor, oxygen and nitrogen. If hot enough, the gases may become ionized to produce plasma. Depending on the substances and any impurities, the color of the flame and the intensity will differ.

So, it is a chemical process in which oxidation is taken place rapidly of material and then it releases heat, light and products. Further, at a certain point visible flame is produced. So initially oxidation is happening and then heat and then flame.

2) A rapid, persistent chemical change that releases heat and light and is accompanied by flame, especially the exothermic oxidation of a combustible substance.

3) This is same as above, but flame is produced quite ahead along with reaction.

4) A state, process, or instance of combustion in which fuel or other material is ignited and combined with oxygen, giving off light, heat, and flame.

As per this slight variant is happening that combustion happens and fuel is ignited with existing oxygen, which radiates light, heat and flame.

5) A process in which substances combine chemically with oxygen from the air and typically give out bright light, heat, and smoke; combustion or burning.

As per this chemical and oxygen processed and gives output.

6) The state of burning that produces flames that send out heat and light, and might produce smoke.

In this it is stated as just burning, producing flame.

7) A combustion reaction that results in heat and flames; sometimes purposefully and beneficially (for cooking purposes, or for heating purposes), but sometimes unintentionally (a spark which starts a wildfire, or burns down a house).

   As per this, again combustion happening and further results are seen.

8) The state of combustion in which inflammable material burns, producing heat, flames, and often smoke.

   This definition is almost same as mentioned above.

9) A state, process, or instance of combustion in which fuel or other material is ignited and combined with oxygen, giving off light, heat, and flame.

   This definition is almost same as mentioned above.

It can be seen that fire is process of combustion and during that process there is release of heat, light and flame.

These are dictionary meaning of fire and are more or less is same in all dictionaries except few changes as-

- Fire as rapid persistence chemical change accompanied by flame, oxidation of combustible substance,

- Process in which fuel or other material is ignited,

- In process, oxygen is getting added from air,

- Intentional combustion, such as cooking or heating purpose,

- Unintentional combustion such as spark or wild fires, process in which fuel or other materials are ignited along with air and emittance of is light, flame, smoke, burning.

Dictionary meaning is derivatives of various theories and fire is hovering around combustion, oxidation, fuel ignition, spark etc. and flame is produced emitting light, heat, smoke and gases.

HISTORY OF FIRE

Fire is a natural reaction that didn't need to be invented, although there are or could be a lot of mythological stories. The earliest creatures that predated human beings were probably well aware of fire. When lightning would have struck, then there must be fear in their mind or could be amazed or be intrigued and curiosity would have stretched their thinking towards this natural phenomenon. Sometimes burning of forest due to lightning could have been strange initially but must have been thought process for them.

Exactly who first learned to understand, create and control fire? And when did that happen? These are questions that have no definite answers. In fact, they're some of the most asked questions that scientists still study today and archeologist try to find the answer in which era this must have happened. But with the present evidences are leading to some clues and definite answers although not precise timing.

The earliest humans must have terrified of fire just as animals were. But they had the intelligence to recognize that they could use fire for a variety of purposes. Fire provided warmth and light and kept wild animals away at night. Fire was useful in hunting. Hunters with torches could drive a herd of animals over the edge of a cliff. Even fire was useful for cooking.

Although it's impossible to know exactly who used fire or how they used it, experts believe these sites show that the early ancestors of human beings used controlled fire well over a million years ago. Evidence at some of the sites indicates that the use of fire could date back almost two million years. As cultural growth of human and progress happened the intimate connection of fire must have taken place.

Today, many scientists believe that the controlled use of fire was likely first achieved by an ancient human ancestor known as **Homo erectus** during the Early Stone Age. Archeologists uncovered evidence of what they believe to be the controlled use of fire in Wonderwerk cave in South Africa, as well as the Lake Turkana region of Kenya.

Early humans started a fire and controlled the use of fire. If early humans controlled it, how did they start a fire? We do not have firm answers, but they might have used pieces of flint stones banged together to create sparks. They might have rubbed two sticks together, generating enough heat to start a blaze. Conditions of these sticks had to be ideal for a fire. Learning to make and control fire was most likely one of the earliest discoveries made by pre-humans that walked upright on two legs.

Through the centuries, there has been such an intimate connection of fire with the cultural growth of humanity that whatever relates to the antiquity of fire is important in tracing the history of early progress. Because as years passed, the early stove, early chimney, ancient light came and there after leap and bound technological advancement happened in next year.

However, for us taking forward the topic of fire, conclusion can be drawn that <u>fire was natural phenomenon and subsequently human being understood the same and started the fire, controlled the same and utilize the for our growth.</u>

But the situation has now made by us that we have done growth and due to our mistakes fires are happening and although we are using fire in daily use in case of accidents the same entity has become unwanted due to mistakes of human being and cursing this fire instead taking blame on us.

Fire is god gift and it is not the mistake of fire but of human beings who are culminating the fire.

LIST OF RECENT FIRES

ELECTRICAL FIRE INCIDENCES IN RECENT PAST IN INDIA DUE TO SHORTCIRCUIT		
Sr. No.	Date	Incidence
1	09-02-2021	A major fire broke out at the at the power loom unit located at Ashwani Kumar Road, Surat. The fire broke out on the ground floor due to short circuit in the electric metre box.
2	08-02-2021	Major fire at Guwahati shopping mall. The alcohol stock of the bar caught fire, which is doubted to be started from the electrical panel of the outlet,
3	26-01-2021	Sukharam Textile park at Kadodara Surat. Fire could have occurred due to the short circuit in electricity line due to increase in load.
4	22-01-2021	Earlier in the morning, a minor fire broke out in the meter board at the Institution of Engineers Building at ITO.
5	21-01-2021	Fire incident at Serum factory, Pune. Initial fire was caused by ongoing welding work at the site.
6	09-01-21	Fire in special newborn care unit of the district hospital in Maharashtra's Bhandara and ten infants were killed.
7	19-01-21	Serum Institute Fire due to welding work at the site.
8	16-12-2020	In Bikaner high tension wire fell on truck.

ELECTRICAL FIRE INCIDENCES IN RECENT PAST IN INDIA DUE TO SHORTCIRCUIT		
Sr. No.	Date	Incidence
9	08-12-2020	Fire broke out at Hotel Nyay Mandir Bharuch. Fire could erupt due to short circuit incident.
10	06-12-2020	Major fire breaks out at Shyam Shikar Complex in Ahmedabad.
11	27-11-2020	Fire in Uday Shivanand Hospital Rajkot broke out in intensive care unit allegedly because of a short circuit in the electrical equipments.
12	15-11-2020	Shops gutted in fire in Ambala Haryana. Fire is suspected due to short circuit and then fire crackers might have been further reason.
13	12-11-2020	Chemical company at Dhatav Ind estate MIDC Roha, Raigad. Fire broke out due to short circuit at company's storage plant.
14	23-10-20	Fire in Mumbai three floors mall housing around 400 shops on each floor and was doused after 56 hrs. The reason was short circuit in a shop on the second floor.
15	05-10-2020	Major fire broke at paint factory at Kanpur. The fire has started due to short circuit but exact cause to ascertain.
16	12-08-2020	Fire incident in Parliament Annexe building. Suspected fire due to short circuit.
17	04-08-2020	Over 100 shops were gutted as major fire ravaged a market in Baripur town in west Bengal's South 24 Pargana district. A preliminary investigation found that the blaze was caused by an electrical short circuit.
18	04-06-2020	A major fire broke out in the plastic packaging unit located in the diamond park of GIDC Surat. The unit is making packaging material for the chemical companies and that the fire broke out due to short circuit.

ELECTRICAL FIRE INCIDENCES IN RECENT PAST IN INDIA DUE TO SHORTCIRCUIT		
Sr. No.	**Date**	**Incidence**
19	28-05-2020	Fire broke out at Hotel Fortune at Dhobi Talao Mumbai. The blaze was confined to cables in electrical duct up to third floor and false ceiling in lobby.
20	08-11-2019	Fire in KEM hospital Govt. Mumbai, Maharashtra Night Pediatric ward due to Short circuit
21	08-03-2019	SCB Medical college and Hospital, Govt. Cuttack, Orissa, fire took place in Night Pathology Department. Short circuit as AC were left on during the night.
22	07-02-2019	Fire in Metro Hospital and Heart Institute Pvt. Noida, Uttar Pradesh as Short circuit in water heater inside recovery area.
23	31-12-2018	Calcutta Medical College and Hospital Govt. Kolkata, West Bengal , haematology department Short circuit in a refrigerator in haematology deptt.
24	20-12-2018	Fire in ESIC Kamgar Hospital Govt. Suburban Andheri, Mumbai, Maharashtra due to Short circuit in AC
25	20-11-2018	Chittaranjan National Cancer Institute Govt. South Kolkata, West Bengal Afternoon Genetic research wing fire due to Short circuit in AC
26	12-11-2018	Calcutta School of Tropical Medicine Govt. Calcutta, West Bengal fire due to Short circuit in AC
27	18-10-2016	SUM Hospital, Shiksha 'O' Anusandha n Univ. Govt. Bhubaneswar, Orissa fire due to Short circuit

ELECTRICAL FIRE INCIDENCES IN RECENT PAST IN INDIA DUE TO SHORTCIRCUIT		
Sr. No.	Date	Incidence
28	21-09-2016	Safdarjung Hospital Govt. New Delhi Afternoon Ground floor, near Casualty Short circuit from meter box
29	31-05-2016	SCB Medical College Govt. Cuttack, Orissa Afternoon 2 nd floor Duty room fire due to Short circuit in AC
30	14-02-2016	The fire broke out at 'Make in India' event in Mumbai was reportedly caused due to electrical short circuit. During examination the officials found that an electrical spark had come in contact with the inflammable materials under the stage.
31	29-11-2015	Sardar Vallabhai Patel Postgraduate Institute of Paediatrics Orissa fire due to Short circuit in Electrical warmer
32	16-10-2015	Acharya Harihar Regional Cancer Centre Govt. Cuttack, Orissa fire due to Short circuit in AC
33	13-01-2013	PBM Govt. Bikaner, Rajasthan fire in ICU Short circuit in AC
34	06-09-2012	NEW DELHI: Ghazipur slaughterhouse fire broke out in the air-conditioning.
35	06-09-2012	JAIPUR: fire broke out at a multi-storey, fire was caused by short-circuit.
36	05-09-2012	MARGAO: A fire broke out in a shop selling electrical goods, short circuit may have caused the fire.
37	04-09-2012	Fire mishap at Apollo Tyres, Kalamassery, Short circuit in the electrical devices is supposed to be the reason for the accident.

ELECTRICAL FIRE INCIDENCES IN RECENT PAST IN INDIA DUE TO SHORTCIRCUIT		
Sr. No.	Date	Incidence
38	01-09-2012	Fire broke out in a godown in Burail, Chandigarh due to Short-circuit in the electrical wires.
39	23-08-2012	A fire broke out at the Cumballa Hill. The fire was caused by a short-circuit in the air-conditioning duct.
40	22-08-2012	A fire broke in a four-storeyed building, Anjali at Opera House in the meter room in the basement, and was probably due to short-circuit.
41	20-08-2012	Fire broke out in hostel of School in DLF V Gurgaon due to an electric short circuit in an air-conditioner.
42	17-08-2012	SCB Medical College and Hospital Cuttak fire. Short circuit caused a fire in the ICU of the neurosurgery department.
43	13-08-2012	A warehouse was gutted in fire due to short-circuit in Kubernagar Ahmedabad.
44	04-08-2012	Fire ravaged a three-star hotel in Barbil mining area Keonjhar. The fire might have resulted from a short circuit in the ground floor.
45	31-07-2012	In Surat electric transformer of DGVCL exploded on Sunday evening.
46	26-07-2012	Car showroom at Udhna Darwaza, Surat fire started at the showroom due to short circuit in the AC system.
47	24-06-2012	A short-circuit in a server room of Mantralaya's main building Mumbai was believed to have caused the fire.
48	07-03-2011	Fire broke out at a saree shop, Kanpur. Short circuit was probably the cause of the fire.

ELECTRICAL FIRE INCIDENCES IN RECENT PAST IN INDIA DUE TO SHORTCIRCUIT		
Sr. No.	**Date**	**Incidence**
49	18-11-2010	Fire broke out on multi-storied building Kolkatta from the electric meter box, probably because of a short circuit.
50	17-11-2009	Fire engulfed two shops, Vasco due to short circuit.
51	29-04-2009	A shop selling plastic items in Belgaum was gutted in a fire that occurred due to a short circuit in Khadi Bazaar.
52	13-06-1997	Uphaar film theatre New Delhi fire is one of worst tragedies. A small fire had broken out in an electric transformer which later exploded in flames.
53	23-12-1995	Fire broke out at a school in Haryana's Mandi Dabwali town. The fire was caused by a short circuit in an electric generator during a school event.

REFERENCES

1) Kulkarni Ajit, Fire due to Electricity, Mumbai, Indian Electrical & Electronics Manufacturers' Association, 2012

2) Wikipedia. 2021. "Fire."Last modified November 21, 2021. http://en.wikipedia.org/wiki/Fire

3) The Free Dictionary by Farlex. Retrieved November 28 2021. http://www.thefreedictionary.com/fire

4) Dictionary.com Unabridged. Random House Unabridged Dictionary. 2021. http://dictionary.reference.com/browse/fire

5) Oxford University Press. Retrieved in November 2021.http://www.oxforddictionaries.com/definition/english/fire

6) Cambridge University Press. Retrieved in November 2021. http://dictionary.cambridge.org/dictionary/british/fire

7) http://www.businessdictionary.com/definition/fire.html

8) Reverso-Softissimo, 2021. http://dictionnaire.reverso.net/anglais-definition/fire%20away

9) Sandbox & Co. Infroplease, retrieved in November 2021. http://dictionary.infoplease.com/fire

10) Independence Hall Association in Philadelphia. ushistory.org, Accessed in November 2021. http://www.ushistory.org/civ/2d.asp

11) International Association of Fire and Rescue Services, Slovenia, Prof. Dr. Nikolay Brushlinsky, Prof. Dr. Sergei Sokolov, Dr. Ing. Peter Wagner, Marty Ahrens, Dr. Jury Kolomietz.

    https://www.ctif.org/sites/default/files/2018/06/CTIF_Report23_World_Fire_Statistics_2018_vs_2_0.pdf

12) National Crime Records Bureau, New Delhi, India,

    https://ncrb.gov.in/sites/default/files/adsi_reports_previous_year/table-1.10_0.pdf

13) Hindustan Times, Nov 2017, https://www.hindustantimes.com/mumbai-news/short-circuit-has-caused-80-of-mumbai-fires-over-past-three-years-says-fire-brigade/story-VKbmOh1xzYLytukoEvzwnM.html

14) The Times Of India, July 2019, https://timesofindia.indiatimes.com/city/allahabad/allahabad-

short-circuits-cause-70-percent-fire-mishaps/
articleshow/70042485.cms

15) The Hindu, Feb 2019 https://www.thehindu.com/
life-and-style/homes-and-gardens/why-safe-wiring-
is-a-must/article26330183.ece

16) The Indian Express, Jan 2019 https://indianexpress.
com/article/cities/mumbai/mumbai-70-to-80-fires-
due-to-faulty-electrics-says-bmc-5543949/

17) Department of Delhi Fire Services May 2018, https://
dfs.delhigovt.nic.in/content/fire-safety-precautions-
against-electricity

18) National Building Code of India 2016, published by
Bureau of Indian Standards, New Delhi, India, National
Building Code 2016

19) National Electrical Code 2011, published by Bureau
of Indian Standards, New Delhi, India, NEC

20) Central Electricity Authority Regulations, 2010 by
Central Electricity Authority of India under Indian
Electricity Act, 2003 - CEAR Rules

21) International Standard, International Electrotechnical
Commission, Switzerland, IEC 60364

22) International Standard, International Electrotechnical
Commission, Switzerland, IEC 61439

23) Indian Standards, Published by Bureau of Indian
Standards, New Delhi, India, IS 2309, IS 3034, IS

1646, IS 732, IS 694, IS 5571- 5572, IS60947, IS5216, IS61439, IS12640, IS3043, IS62305

24) The National Geographic Society, Washington https://www.nationalgeographic.com/environment/natural-disasters/lightning

25) OBO BETTERMANN GmbH & Co. KG, 58694 Menden, Germany https://obo.com.cn/media/Lightning_protection_guide.pdf

26) IndiaSpend, Khandekar Nivedita,18 Aug, 2020 https://www.indiaspend.com/why-hundreds-die-of-lightning-strikes-despite-new-technology-programmes/#:~:text=More%20than%2040%2C000%20deaths%20in,see%20chart)%20on%20accidental%20deaths.

27) Conditioned Air Solution, Huntsville, Alabama 35806

https://www.conditionedairsolutions.com/protecting-your-home-from-an-air-conditioner-fire/#:~:text=Air%20conditioner%20fires%20can%20have,so%20can%20loose%20electrical%20connections.&text=This%20can%20cause%20even%20a,to%20overheat%20and%20catch%20fire.

28) India Today, Sept 2019

https://www.indiatoday.in/mail-today/story/watch-out-your-ac-can-start-a-massive-fire-1595177-2019-09-04

29) Ajit Kulkarni Consultants Pvt. Ltd. Testing Format AKC ISO Documents

30) Occupational Health & Safety, Recognizing and Mitigating Static Electricity Hazards, By Karen D. Hamel, Jun 01, 2020

https://ohsonline.com/Articles/2020/06/01/Recognizing-and-Mitigating-Static-Electricity-Hazards.aspx

31) SANKEN ELECTRIC CO., LTD., Measures for Electrostatic Discharge (ESD)

https://www.semicon.sanken-ele.co.jp/en/support/reliability/4-9.html

32) NFPA Public Education Division, Quincy, MA 02169, Hoarding and Fire : Reducing the Risk, https://www.nfpa.org/~/media/files/public-education/resources/safety-tip-sheets/hoardingtipsheet.pdf?la=en

33) Inter Fire Online, Presented by: Dr V Babrauskas at the 7th international Fire & Materials conference, 2001, San Francisco, USA, http://www.interfire.org/features/electric_wiring_faults.asp

34) Electrical Installation WiKi. This wiki is a collaborative platform, brought by Schneider Electric, 20 December 2019

https://www.electricalinstallation.org/enwiki/Quality_and_safety_of_an_electrical_installation#Initial_testing_of_an_installation

35) Megger, Technical Library https://megger.com//support/technical-library

36) RR Kabel, Wires & Cables, Construction and Building Range, https://www.rrkabel.com/construction-building-range/

37) Electrical Engineering Portal, 15 Bad Situations that may lead to catastrophic explosion of a capacitor Bank, By Edvard, 23 Dec 2018, https://electrical-engineering-portal.com/capacitor-bank-catastrophic-explosion

38) American Welding Society, 8669 NW 36 Street, # 130 Miami, Florida, FIVE TIPS FOR PREVENTING WELDING FIRES, AUGUST 31, 2016

https://awo.aws.org/2016/08/five-tips-for-preventing-welding-fires/

39) Welder, Preventing welding-related fires, By Vicki Bell, October 12, 2004, https://www.thefabricator.com/thewelder/article/safety/preventing-welding-related-fires

40) ResearchGate, A study on reported fire incidents in major hospitals of India, International Journal of Community Medicine and Public Health 7(10):3896, DOI:10.18203/2394-6040.ijcmph20204351, September 2020, Shyam Siddharth Rao Patharla, Souri Reddy Pyreddy, Shilpa N. Panthagani.

https://www.researchgate.net/publication/345661050_A_study_on_reported_fire_incidents_in_major_hospitals_of_India

41) India.com, Make in India fire: Stage caught fire due to electrical short circuit, February 24, 2016

https://www.india.com/news/india/make-in-india-fire-stage-caught-fire-due-to-electrical-short-circuit-says-report-977207/

42) The Times of India, Major fire breaks out at GST Bhavan in Mumbai's Byculla, Richa Pinto & Vijay V Singh / TNN / Updated: Feb 18, 2020

https://timesofindia.indiatimes.com/city/mumbai/mumbai-fire-breaks-out-at-gst-bhavan-in-byculla-east/articleshow/74171725.cms

43) Havells India Ltd, LT/HT Power and Control Cables, Catalogue 2016 https://www.havells.com/content/dam/havells/brouchers/Industrial%20Cable/Cable%20Catalogue-2016.pdf

44) Eaton, Switzerland, Publication No. BR003012EN, January 2021, AFDD+ Range Broucher, https://www.eaton.com/content/dam/eaton/markets/residential/fire-safety/documents/fire-safety-afdd-range-brochure.pdf

45) Fike Corporation, Types of Fire Suppression Systems, 101 ARTICLES, FEBRUARY 15, 2019 https://www.

fike.com/knowledge-center/101/types-of-fire-suppression-systems/

46) CTR Manufacturing Industries Private Ltd., Pune 411014, India, EXPLOSION PREVENTION AND FIRE EXTINGUISHING SYSTEM FOR TRANSFORMERS AND REACTORS https://ctr.in/product/transformer-explosion-prevention-and-fire-extinguishing-system-for-transformers-and-reactors/

47) Cholamandalam MS Risk Services Ltd, Chennai – 600031, India, ARC FLASH

http://www.cholarisk.com/services/electrical-safety/arc-flash/

48) Schneider Electric, 2021, Arc Flash Protection, https://www.se.com/in/en/work/services/field-services/electrical-distribution/optimise/arc-flash-solutions/

49) Electrical Installation Wiki, Protection against electrical fire risks, this wiki is a collaborative platform, brought by Schneider Electric 28 Aug 2020

https://www.electrical-installation.org/enwiki/Protection_against_electrical_fire_risks

50) AquaMist, Fire Suppression Products, a business unit of Johnson Controls, Water Mist Fire Protection Solutions. 2021, https://tycoaquamist.com/

51) Xtralis, New Delhi India,2021, VESDA Aspirating Smoke Detection https://xtralis.com

52) Schneider Electric, Electrical fire prevention guide, 2021

https://go.schneider-electric.com/WW_201907_Electrical-Fire-Prevention-Guide-Content_EA-LP-EN.html?_ga=2.230587736.533401090.1622446301-1153058678.1467626963

53) Schneider Electric, White Paper on AFDD, How Arc Fault Detection Devices Minimize Electrical Fire Threats, Jean-Francois Rey, Pierre Dominique Socquet-Clerc, Simon Tian and Zakaria Belhaja

https://download.schneider-electric.com/files?p_enDocType=White+Paper&p_File_Name=EDCED117020EN.pdf&p_Doc_Ref=EDCED117020EN

54) Schneider Electric, White Paper HeatTag, New gas and particle sensing technology detects cables overheating in LV equipment, Abderrahmane Agnaou and Mathieu Guillot

https://go.schneider-electric.com/WW_202105_WP-Heat-Tag_MF-LP.html?source=Content&sDetail=WP-Heat-Tag_WW

55) The Times of India, August 2014, digital image

https://timesofindia.indiatimes.com/city/mumbai/mumbai-has-most-short-circuit-fires-and-deaths-in-india/articleshow/40026142.cms

56) Indian Standard Code for Practice of Earthing, published by Bureau of Indian Standards, New Delhi, India, digital images

57) All Images of detection and Fire panels are from MIRCOM

58) https://www.legrand.nl/sites/default/files/2017-11/EXB15004_IEC61439_June_2015.pdf

59) https://www.se.com/myschneider/documentsDownloadCenterDetail/in/en/ENVPEP1401004EN

60) https://www.lntebg.in/media/41926/ti-small-catalouge_sp50685r3.pdf